MATHEMATICAL MODELING OF BIOLOGICAL SYSTEMS

MATHEMATICAL MODELING OF BIOLOGICAL SYSTEMS— AN INTRODUCTORY GUIDEBOOK

HARVEY J. GOLD
Biomathematics Program
Department of Statistics
North Carolina State University
Raleigh

A WILEY-INTERSCIENCE PUBLICATION

John Wiley & Sons, New York ● London ● Sydney ● Toronto

Copyright © 1977 by John Wiley & Sons, Inc.

All rights reserved. Published simultaneously in Canada.

No part of this book may be reproduced by any means, nor transmitted, nor translated into a machine language without the written permission of the publisher.

Library of Congress Cataloging in Publication Data:

Gold, Harvey J   1932-
   Mathematical modeling of biological systems.

   "A Wiley-Interscience publication."
   Bibliography: p.
   Includes indexes.
   1. Biology--Mathematical models.   I.   Title.
QH323.5.G64   574'.01'84   77-8193
ISBN 0-471-02092-3

Printed in the United States of America

10 9 8 7 6 5 4 3 2 1

PREFACE

This book is intended for biologists and biology students who have an interest in mathematical modeling but who have a limited background in mathematics. Its purpose is to assist the student in understanding how biological concepts may be formulated in mathematical terms and how these formulations may be used in biological research. The book is based on notes that I have used for several years as a classroom text in a course for beginning graduate students in biology. Comments from these students encouraged the not inconsiderable work of turning the notes into a textbook. I have sought to make it the type of book that I would like to have had available some fourteen years ago when, as an experimental biologist, I first became interested in mathematical models.

Biological systems are among the most complex studied by modern science. As our experimental knowledge of these systems becomes more detailed, it becomes increasingly difficult to organize this knowledge without the "bookkeeping" and deductive methods of mathematics. The research literature of experimental biology bears witness to the increasing use of mathematical methods. Nevertheless, biologists often lack the training to make proper use of these methods. More to the point of this book is the fact that the biologist is frequently unable to grasp the biological relevance of mathematical models encountered in the literature or to critically evaluate that relevance.

A number of texts are available that treat the application of mathematics in biology. However, the modeling process is rarely examined in sufficient depth so that the reader can see the relation between an individual mathematical expression and his or her "biological intuition." The reader is commonly advised that the relation between a certain pair of variables obeys "the following" equation or that the rate of change of a population size obeys "the following" rate law. The logical and mathematical consequences of the stated equation are then explored while the reader is left to guess why such an equation should be used, what *physical and biological assumptions* are implied by the use of the equation, and when the equation should not be used.

The point of view taken in this book is that certain *physical* characteristics of a system lead to particular types of *mathematical* descriptions. An example is the association of hyperbolic rate laws (such as are found in descriptions of enzyme kinetics) with competition for a limited renewable resource.

Mathematical prerequisites for the book are minimal, and a brief review is provided in the appendix. Some familarity with the concepts of calculus is necessary, but the demands made upon the reader's calculus background are not great. In spite of this, the range of mathematical topics discussed is wide and includes some topics normally reserved for advanced level instruction in mathematics. As already stated, the intent is to teach the process by which the properties of systems can be captured in the framework of mathematical structures. The primary concern is, therefore, with the conceptual content of the relevant mathematical structures rather than with the manipulation of equations. A significant portion of this conceptual content can be made comprehensible to a student whose formal mathematical background includes only an elementary introduction to calculus, providing the student has sufficient grounding in areas of biology (or of general science) to which the concepts are applied.

Examples used in the text may be broadly divided into two kinds. Prototype systems are used for the development of basic principles. These examples often involve very simple systems that are treated in great detail so that the principles being developed stand out in bold relief. Examples of more realistic complexity are then used to illustrate the application of the principles. These are drawn from a wide variety of biological areas, selected so as to be understandable to students with general background in biology without an inordinate amount of specialized study. In addition, several examples are developed in more detail in Appendix A.

A certain number of exercises are scattered through the book. Although some of these are problems to which answers can (and are) given "in the back of the book," my favorite type of exercise runs something like: "Apply the principles just discussed to an appropriate system from your own field of biology. Justify the application and elaborate on the consequences." In this spirit, the reader is asked in Chapter 2 to begin the development of a model for a system of his or her own choosing. Many of the subsequent exercises are really hints and prods in the continued development of that model. In a classroom situation, such exercises may form the basis of oral reports and general class discussion. The reaction of most students has been that these exercises constitute an extremely valuable part of the course. The reader who is using this book for self-study is

PREFACE

urged not to neglect them; indeed, it would be best to discuss at least some of them with a knowledgeable associate who would enjoy finding flaws in the reasoning.

Facility in mathematical modeling can come only through the experience of developing one's own models and the critical study of models developed by others. A prime objective of this book is to acquaint the reader with principles that will guide such further exploration; it is in this sense that the term *Guidebook* is used in the title.

It gives me great pleasure to acknowledge my indebtedness to Henry L. Lucas and H. Robert van der Vaart for many long and stimulating discussions about mathematical modeling. My sincere thanks go to Robert Geckler and Timothy F. Allen, who examined the manuscript, for their thoughtful and helpful comments. Portions of the manuscript were also examined by Ronald E. Stinner, Carl W. Dawson, and Charles Martel; my thanks to them for their valuable comments. Special thanks are due to Nancy K. Evans, who typed the manuscript, for her careful and intelligent handling of a seemingly endless succession of drafts. Finally, I wish to thank Shirley Gold for her invaluable editorial advice and for her persistent posing of the all-important question, "But, what does the mathematics *mean*?"

HARVEY J. GOLD

Raleigh, North Carolina
April 1977

CONTENTS

Symbols and Notation, xiii

1. **Introduction**   1

    1.1. Mathematical Models, 2
    1.2. Correlative Models, 3
    1.3. Explanatory Models, 8
    1.4. Scientific Theory and the Role of Mathematics, 10
    1.5. Mathematical Proof and Physical Proof, 12
    1.6. Idealization and Simplification, 14
    1.7. Mathematical Results and Biological Intuition, 15
    1.8. Further Reading, 15

2. **The Modeling Process—An Overview**   18

    2.1. A Simple Public Address System, 19
    2.2. Summary of the Model-Building Protocol, 25
    2.3. Formulation of the Problem, 27
    2.4. Decomposing the System, 27
    2.5. Variables, Parameters, and the State of the System, 31
    2.6. Signal Flow Graphs. Relations Between Variables, 32
    2.7. Determining the Form of the Input–Output Relations, 34

3. **Dimension and Similarity**   39

    3.1. Fundamental Physical Dimensions, 40
    3.2. Simple Algebraic Rules for Dimension, 40
    3.3. The Concepts of Scale and Dimension, 44
    3.4. Apparent Conversion of One Dimension to Another, 48

- 3.5. Dimensional Homogeneity, 49
- 3.6. Similarity, 51
- 3.7. Scale Limitations, 59
- 3.8. Further Reading, 60

4. Probability Models — 61

- 4.1. Probabilities and Weighted Averages, 62
- 4.2. Example from Mendelian Genetics; Definition of the Problem, 64
- 4.3. The Sample Space or Set of Outcomes, 66
- 4.4. Events and Their Probabilities, 68
- 4.5. Simple Counting Rules, 70
- 4.6. Combinations of Independent Events, 77
- 4.7. Conditional Probability, 82
- 4.8. Expected Values, 85
- 4.9. Variability, 88
- 4.10. Summary of Basic Ideas, 92
- 4.11. Example from Mendelian Genetics; Formulation of the Model, 94
- 4.12 Uncertainties in the System: Stochastic versus Deterministic Models, 96

5. Dynamic Processes — 100

- 5.1. Processes with Constant Average Rate, 103
- 5.2. Linear Rate Laws, 114
- 5.3. Behavior of Simple Linear Systems, 117
- 5.4. Characterizing the Rate of Exponential Change, 120
- 5.5. Examples of Simple Linear Rate Processes, 123
- 5.6. Nonhomogeneous Population, 129
- 5.7. Direct Interactions; Chemical Law of Mass Action, 133
- 5.8. Nonhomogeneity with Respect to the Independent Variable, 137
- 5.9. Summarizing Remarks, 144

6. Interacting Dynamic Processes — 145

- 6.1. Isolated Systems and Equilibrium, 146
- 6.2. Open Systems and Steady State, 157
- 6.3. Flow in a General Compartmental Framework, 159
- 6.4. Diffusion and Migration in a "Continuous" System, 164

- 6.5. Chemical Interactions and Equilibrium, 171
- 6.6. Direct Competition Effects in Population Models, 173
- 6.7. More General Interaction Models, 176
- 6.8. Indirect Interaction; Saturation of a Limited Resource, 178
- 6.9. Combining Cooperation and Resource Limitation; The Hill Equation, 182
- 6.10. Some Remarks on Cycling Behavior, 184

7. Feedback Control and Stability of Biological Systems   186

- 7.1. Steady States and Equilibrium, 186
- 7.2. Steady States and System Stability, 190
- 7.3. Steady States and Observability, 192
- 7.4. Feedback Relations and Homeostasis, 193
- 7.5. Example: Lotka–Volterra Model for Two-Species Prey–Predator System, 200
- 7.6. Sensitivity Analysis, 203
- 7.7. Behavior Away from Steady State, 203
- 7.8. Biological Rhythms, 210

8. Curve Fitting: Estimating the Parameters   213

- 8.1. As Many Parameters As Data Points, 214
- 8.2. What Is "Best"? 215
- 8.3. Finding the Best Parameter Values for Linear Equations, 219
- 8.4. How Good Is the Best Fitting Curve? 225
- 8.5. Random versus Systematic Deviations, 226
- 8.6. Resumé of the General Procedure, 228
- 8.7. A Quick and Approximate Method for Estimating a Good Fitting Curve, 228
- 8.8. Unequal Errors. Weighting of Data, 229
- 8.9. Data Transformation and the Error Structure, 232
- 8.10. Correlations Between Variables, 234
- 8.11. Forced Correlations, 236

9. Computing   240

- 9.1. Inside the Machine, 241
- 9.2. Approximate Calculation of Mathematical Functions, 247
- 9.3. Computer Simulation of Systems, 250
- 9.4. Classification by Computer; Numerical Taxonomy, 256

References 261

Appendix A. Examples 266
Appendix B. Mathematical Review 308
Appendix C. Mathematical Approximation 336
Appendix D. Linearity and the Use of Averages 340
Appendix E. Discussion of Exercises 346

Index 351

SYMBOLS AND NOTATION

When treating several topics under one cover, it is difficult to avoid using the same symbol for several purposes. However, different uses for the same symbol are avoided within a single chapter. In the following list, the numerals in parenthesis are chapter numbers. When a particular use for a symbol is shown without a chapter number, it is meant to indicate a general use for the symbol. In many cases, the specific meaning of a symbol is modified by subscripts or superscripts.

a	arbitrary parameter (1, 8); label for an object (3); label for an outcome (4); stoichiometric coefficient (5, 6)
A	name of variable (2); area (3); specific event (4); specific population (6)
b	arbitrary parameter (1, 8); label for an object (3); stoichiometric coefficient (5, 6)
B	name of variable (2); specific event (4); specific population (6)
c	arbitrary parameter (1, 8); allele designation (4); stoichiometric coefficient (5, 6)
C	name of variable (2); allele designation (4); concentration (5, 6); specific population (6)
CV	coefficient of variation
d	distance (3, 8)
D	diffusion coefficient
e	base of natural logarithms; environment subscript (6)
$\mathcal{E}$	expectation
E	efficiency (2, 7); enzyme (6)
F	feedback fraction (2, 7); progeny generations (4)
g	gravity (3)
G	amplifier gain (2, 7)
H.I.	homeostatic index (7)

xiii

I	inputs; intensity (5)
J	flow (6)
k	parameter, usually positive; shape factor (3)
$k_{i,j}$	rate parameter for transfer from i to j (6)
k_f, k_r	chemical reaction rate constants (6)
K	scale factors (3); Michaelis constant (6)
ln	natural logarithm
L	length (3, used for scale and dimension)
m	mass (3); stoichiometric coefficient (6)
M	mass dimension (3); number of photons (5); type of compound (6)
MS_{dev}	mean square deviation (8)
n	general use to indicate integer
$n_{i,0}$	number in ith population at time t_0
N	generally indicates an integer
$o(x)$	lower order of magnitude; see Appendix C
$\mathcal{O}$	output (2)
p	point (3, 8); probability (4, 5, 8); stoichiometric coefficient (6)
p_r	probability of reaction (5)
P	probability (4, 5, 6); reaction product (6); prey population (7, 9)
q	point (3); local production (6); general coordinate (8)
Q	electrical charge dimension (3); condition (4); production term (6); arbitrary quantity (9)
r	number of possibilities or of populations (4, 5); correlation coefficient (8)
$r_{i,j}$	rate of transfer from i to j
R	arbitrary quantity (9)
RF( )	relative frequency (4)
RMS_{dev}	root mean square deviation (9)
S	scale (3); sample space (4); substrate (6)
ss	designates steady state (6)
t	time
T	time dimension; total
u	number of parameters (8)
v	reaction velocity (6)
V	volume (3, 6); an event (4); vegetation (9)
V_{max}	maximum velocity
Var	variance (4, 5)
w	weight of an observation (8)
W	weight (3); number of ways of doing a specific task (4)
x	value of a variable or coordinate

SYMBOLS AND NOTATIONS

X	name of a variable or direction; amount in specified compartment (6)
y	value of a variable or coordinate
$\mathbf{y}$	vector of y values (8)
$\hat{y}$	value predicted on basis of model equation (8)
$\hat{\mathbf{y}}$	vector of predicted values (8)
Y	name of a variable
z	value of a variable or coordinate
Z	name of a variable
α	intrinsic rate of increase (7, 9)
β	interaction parameter (7, 9)
γ	absorption coefficient (5)
δ	time delay (2, 7)
Δ	finite change (5, 6)
ε	error (8)
η	correlation ratio
θ	temperature dimension (3); threshold constant (5)
κ	arbitrary parameter
$\kappa_{i,j}$	interaction parameter (6)
λ	probability of a specific encounter (5)
Π	predator population (7, 9)

MATHEMATICAL MODELING OF BIOLOGICAL SYSTEMS

CHAPTER ONE
INTRODUCTION

*Our choicest plans
have fallen through
Our airiest castles
tumbled over,
Because of lines
we neatly drew
And later
neatly stumbled over.*
Grooks
Piet Hein

The basic *premise* of this book is that mathematical procedures are useful, and sometimes necessary, for the description and understanding of biological systems. The primary *objective* is to help the student understand the relationship between the biological system under study and its mathematical description.

Mathematics may be of service to biology at several conceptually different levels. At one level, it is simply a *tool* playing a role somewhat analogous to that of a spade or of an analytical balance. At this level, it may help get an answer to a question, but the answer can be discussed independently of the tools used. For example, mathematical and statistical analyses may be used to show that some drug is effective against a particular disease, to discover the sequence of chemical reactions in a metabolic pathway, or to find out about the migration habits of birds. In such cases, the mathematics may be thought of as an aid in finding "nonmathematical answers" to "nonmathematical questions."

At a second level, mathematics enters the discussion when we want to

---

Doubleday & Co., Inc., Copyright 1966, *ASPILA SA*. Used by permission of the Author.

know how effective the drug is or how fast the metabolic pathway functions or how fast the birds migrate. In such cases, mathematics is not only a tool, but it is also visible as an integral part of the question and the answer.

Finally, mathematics may become part of the fabric of the answer, even though the question seems to be a nonmathematical one. This is especially so when we are interested in the *relationships* between the parts of a complex system. Appendix A gives several examples.

A mathematical description of a real world system is often referred to as a mathematical model. In this chapter, we take a look at some of the ways in which the term *"model"* is used and do a little philosophizing about the relation between experimental work, theory building, and mathematical modeling. In other words, Chapter 1 develops the conceptual framework in which we operate. The next chapter takes an initial look at the mechanics of model building.

1.1. MATHEMATICAL MODELS

The word "model" is used here with very much the same meaning as its every day meaning. We may say that some object (call it object M) is a model of another object (object S), if the following conditions hold:

i. There is some collection of components of M, each of which corresponds to a component of S;
ii. For at least some relationships, the relation between the components of M is analogous to that between the corresponding components of S.

So, for example, the features of a marionette are intended to correspond to the features of the human being that it models, and certain relations between the features are the same in both. In the same way, an architect's blueprint is a model for a finished building.

A mathematical equation consists simply of a collection of symbols, some of which stand for variables (quantities with number and dimension) and some of which stand for operations on these variables (such as addition, multiplication, and differentiation). If the variables can be associated with physical entities for a given real world system and if the relation between the variables in the mathematical expression is analogous to the relation between the corresponding physical entities, then we may say that the mathematical expression is a model for the real world system.

Conditions i and ii are a *definition* for the word *"model"* in the sense that

1.2 CORRELATIVE MODELS

anything that has these two properties qualifies as a model and anything that lacks one or the other of them does not. Notice, however, that this definition does not require the model M to be an exact duplicate of object S. That is, condition i does not require *every* component of M to correspond to a component of S or vice versa. Nor does condition ii require that every relationship between components of S be mirrored by an analogous relationship in the model. For example, the ratio of arm length to leg length may the same for the marionette as for the human, but the difference between arm length and leg length is not normally the same for the two. Furthermore, the mechanism by which the human moves has no analogy in the mechanism by which the marionette moves. It should be clear that the two objects cannot correspond to each other in every detail unless they are identical objects; in which case, the concept of model looses its usefulness.

In constructing a model, one of the first jobs is, therefore, to decide which characteristics of the object or system of interest are going to be represented in the model. In order to make such decisions, it is necessary that the *purpose of making the model be defined as clearly as possible.*

For our purposes, we may divide mathematical models into two broad types that I call correlative and explanatory models.[1] They are discussed separately in the next two sections.

1.2. CORRELATIVE MODELS

A correlative model is one that is required only to reflect an observed relation between two (or more) variables. Its primary purpose is to describe and to summarize that relationship, usually so that the relation may be verified and then used as a basis for prediction and control.

In constructing such a model (or any mathematical model), we must start with data gathered either from experimentation or from field observations. The "data" comprise whatever we think is relevant. Unfortunately, we don't always know what is "relevant." Even in the chemical laboratory, where everything is generally supposed to be very carefully controlled, there are cases known in which the length of a technician's coffee break, or the precise manner in which the glassware was cleaned, had a great deal to do with the outcome of the experiment. Such things are rarely recorded.

---

[1] A variety of definitions for the idea of model, as well as schemes for classifying models, have been discussed in the literature. The distinction made here between correlative and explanatory models is conceptually similar, though not identical, to the distinction made by Beckner (1959) between nonexplanatory and explanatory models.

EXERCISE 1.1. Describe some experiment in your own field of biology that you have either done or read about.
a. What things were measured?
b. What things were controlled?
c. What things may have been relevant but were neither measured nor controlled?

After examining what is thought to be relevant, a relation may be hypothesized between two variables or among a group of variables; the relation may then be stated in mathematical form. It is then tested to see how closely it agrees with the data. If the agreement is not close enough, a different relation may be tried. Normally we wish to use the simplest mathematical expression that gives sufficient agreement with the data. The process may be pictured as diagrammed on the left side of Figure 1.1.

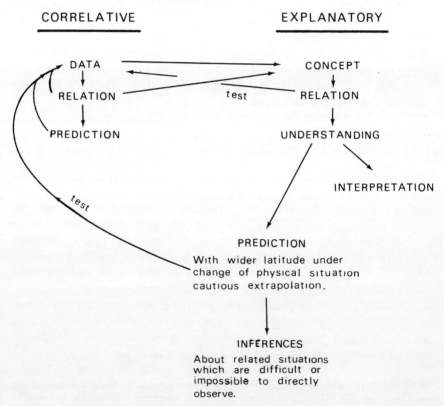

Figure 1.1. Representation of the model-building process. Note that "test" arrows carry one back to the data at every opportunity.

1.2 CORRELATIVE MODELS

1.2a. Choosing the "Right" Description

A clue to the kind of mathematical relation to try can often be obtained by simply plotting the data on graph paper and observing the *qualitative* characteristics of the relation. That is, suppose x and y are two variables that have been measured (amount of fertilizer applied and yield of corn, substrate concentration and the rate of an enzymic reaction, etc.). If a plot of y against x suggests a straight line that goes through the origin, then a reasonable relation between x and y might be a direct proportionality,

$$y = bx \qquad (1.1)$$

where b is some appropriately chosen parameter.[2] If it suggests a straight line that does not go through the origin, we would want the more general relation (lines i and ii, Figure 1.2),

$$y = a + bx \qquad (1.2)$$

If the relation shows curvature of some type (curves iii and iv in Figure 1.2), then we would need a more involved expression, perhaps something of the form,

$$y = a + bx + cx^2 \qquad (1.3)$$

The curvature of a line drawn according to equation (1.3) depends on the relative magnitudes of the parameters b and c.

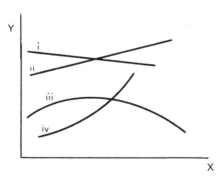

Figure 1.2. Some types of behavior that may be exhibited by curves drawn according to equation (1.3).

[2]The question of how to choose the parameter is discussed in Chapter 8.

EXERCISE 1.2. A good deal of our discussion deals with the description of the qualitative behavior of biological systems. Prove to yourself that a line drawn according to equation (1.3) can have the qualitative appearance of any of the curves in Figure 1.2 by finding values for parameters a, b, and c that give the two different behaviors. Do the curves of Figure 1.2 exhaust all possibilities insofar as the qualitative behavior of (1.3) is concerned? Formulate a concept of what is meant by the "qualitative" appearance of a curve or the "qualitative" behavior of an equation.

EXERCISE 1.3. Cite a pair of variables from your own field of biology that you think would demonstrate each of the kinds of relations shown in Figure 1.2.

1.2b. Nonuniqueness of the Description; Interpolation Between Data Points

Equations (1.1), (1.2), and (1.3) are a few simple examples. In fact, if we are skillful enough and have no reservation about making the mathematical expression exceedingly complex, we can pretty generally get an expression that exhibits whatever kind of twisting, bending behavior we want it to. That is, if the dots in Figure 1.3 represent actual data points, we could, in principle, find an equation that would give the dashed line, which passes through each of the data points, and so reproduces the data exactly. One trouble with actually doing so is illustrated by the dotted line in Figure 1.3. In principle, we could also find an equation that produces this line, and it also passes through every data point, and so exactly agrees with the data. The dashed and dotted lines, however, do very different things in between the data points. Indeed, we could find an infinite number of different curves, each of which passes through all the data points and each of which does something different between them. This dilemma is generally resolved by invoking one of the most basic tenets of science: Use the simplest explanation that fits the facts. In the present context, we would say to use the simplest curve that shows sufficient agreement with the data points. In determining what sort of agreement is sufficient, we must have some estimate of the amount of error associated with the data points; it is nonsense to require that the agreement between the curve and the data points be better than the experimental error. Taking this into account, the data points in Figure 1.3 may be adequately represented by the smooth curve, which actually passes through none of them. Moreover, for some purposes, we do not require the ultimate in accuracy so that the straight line, which is not the best representation of the data, may be adequate.

We see, therefore, that a fundamental difficulty in choosing a mathe-

1.2 CORRELATIVE MODELS

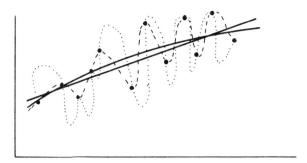

Figure 1.3. The dangers of interpolation.

matical description of the relationship between two variables is that data do not determine a unique description. More than one description may be acceptable. The problem is to develop some set of criteria upon which to base a choice. This problem is discussed in more detail in Chapter 8.

1.2c. Nonuniqueness of the Description; Extrapolation Beyond the Data

Another problem related to the nonuniqueness of the description is that of determining what the mathematical expression does outside the range of the data that have in fact been observed. This is illustrated by Figure 1.4, in which the hypothetical data points seem to agree equally well with the

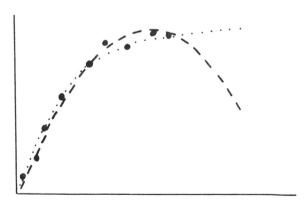

Figure 1.4. The dangers of extrapolation.

two equations,

$$y = 2.7333x - 0.433x^2 \qquad \text{(dashed curve)}$$
$$y = 4.4(1 - e^{-0.86x}) \qquad \text{(dotted curve)}$$

As long as the purpose is to describe the data within the range of observation, either mathematical expression may do nicely. However, this particular set of data offers no hint as to which expression (or another one altogether) should be used outside of this range. Independent knowledge from previous experiments or from the literature may, of course, be helpful. If the data represents a plot of enzyme activity against acidity, we would normally expect behavior similar to that of the dashed curve. If we are dealing with yield of corn plotted against level of nitrogen in the soil, then we would expect behavior similar to that of the dotted curve. If the data are all that we have to go on, however, then to extrapolate beyond the range of observation is to invite disaster.

1.3. EXPLANATORY MODELS

An explanatory model is also required to reflect observed relationships between variables. But, in addition, the structure of the model is required to reflect some concept of the *causal mechanism* that underlies the relation. The purpose is not so much to describe the observed relation as to explain it.[3] In constructing such a model, one generally tries to arrive as soon as possible at an acceptable explanatory concept. The concept is then translated into a mathematical relation, or set of relations, and the predictions of this relation (or set of relations) are compared with the data. This comparison now becomes a test of the concept originally formulated. In this case, disagreements between the model and the experiments cannot be resolved by simply adding convenient mathematical terms to the model. The resolution of the disagreement must involve an analysis of the physical reasons for the disagreement. Often it involves scrapping the whole thing and starting over. The process may be represented as shown in the right-hand portion of Figure 1.1. Once we have attained sufficient agreement with the experimental observations to give us some degree of confidence in our understanding of the system, we may begin to extrapolate

---

[3]The problem of defining the words "cause" and "explanation" has received a great deal of attention in the literature of the philosophy of science. For the reader who wishes to pursue this topic, a few suggestions for further reading are listed at the end of the chapter.

1.3 EXPLANATORY MODELS

cautiously beyond the range of actual experimental observation. As a part of this extrapolation, we may "ask" the model for a prediction of what would happen if we were to modify certain characteristics of the system. These predictions should be tested experimentally as soon as possible. The agreement, or lack of it, then becomes an additional test of our understanding.

As before, we have the problem that an explanation based on a given body of experimental evidence may not be unique. When possible, it is a good idea to develop two (or more) alternative explanatory models that fit the available evidence equally well. Extrapolation often reveals that there is some set of experimental conditions for which the two models disagree, suggesting experiments that would discriminate between the two. Finally, when one finds that a given explanatory model has stood up to a wide variety of tests, he or she may be emboldened to accept the explanation as a working hypothesis and to use it to draw inferences about related systems, under conditions that are either difficult or impossible to examine experimentally. Such inferences may be used as a basis for system control or management or the formulation of public policy (see, for example, the discussion by Cooper, 1976).

EXERCISE 1.4. In every field of science, there are examples of accepted hypotheses (not necessarily of a mathematical nature) used to draw inferences to conditions not amenable to direct experimental tests. Examples are the drawing of inferences about the behavior of individual molecules from the observed behavior of collections of molecules, inferences about the behavior of genes from observed phenotypic behavior, and medical diagnoses on the basis of observed clinical symptoms. The reader should make up a partial list of examples from his own field of biology showing specific applications.

It is important to emphasize that in both correlative and explanatory models, one starts with data (observations), repeatedly returns to the data for the purpose of testing and modifying the model, while using the model to suggest what new data should be gathered. That is, there is an iterative interplay between the experimental work and the modeling work. Indeed, one often finds that the formulation of an explanatory model is frustrated by gaps in the existing body of factual information. Such gaps serve as starting points in the design of new and interesting experiments, and their highlighting may be one of the most valuable results of the modeling activity.

The process of model construction is part of the general process of the building of scientific theory and is used in some form in every field of science. However, the construction of such models takes on added significance in biology because of the enormous complexity of biological systems. This complexity leads to the uncomfortable circumstance that the things we are interested in are often not the things we can directly measure. In order to test our conjectures about the things we are interested in, we must draw some relation between them and the things we *can* measure. The discovery of such relationships is part of the business of the mathematical development of the model.

An interesting example of the interplay between theory and experiment, on a grand scale, is the story of the discovery of DNA as the genetic material. Observation of the transmission of phenotypic characteristics from one generation to another led to the realization that there had to be a parallel transmission of some material substance. Quantitative study of heredity, including the effects of various types of crosses and mutations, built up a fund of information as to the properties that the material substance would have to have. The question then became, "What kind of material substance could exhibit such properties?" In getting the answer to this question, help came from molecular theory (Timofeeff-Ressovsky, Zimmer, and Delbruck, 1935; Schrodinger, 1945). In brief, the answer was that it would have to be something in the nature of a structurally stable supermolecule, capable of exhibiting a long-range *nonperiodic* order. Proteins and nucleic acids were the prime candidates. This background provided motivation (see, for example, Olby, 1974, Sections IV and V) for a great deal of experimental work. The nucleic acids were hailed as "winner" when X-ray crystallography showed DNA to possess a structure that suggested a mechanism for self-replication.

1.4. SCIENTIFIC THEORY AND THE ROLE OF MATHEMATICS

For many biologists, the word "theory" evokes a vision of complicated formulas that can have no genuine relevance to experimental biology. A more accurate interpretation of the word "theory" is simply an understanding of the underlying relations between the individual experimental observations and the "big picture" that cements them together, as the individual pieces of a jig-saw puzzle relate to the picture formed when the pieces have been correctly assembled.

Every experimentalist is also a theoretician to the extent that he or she uses logic and reason in the planning and interpretation of experiments. However, problems develop as the quantity of data multiplies and as the

1.4 SCIENTIFIC THEORY AND THE ROLE OF MATHEMATICS

reasoning process becomes longer and more involved. We begin to require a system of "bookkeeping" to insure that things are being kept straight and some system of symbolic notation so that we can conveniently refer to the same concept as often as we wish. Systems of notation and rules of bookkeeping are the first tools that mathematics offers.

A second category of tools may be referred to as a collection of "prefabricated blocks of logic." The need for these becomes apparent when we find ourselves retracing the same line of reasoning in several different contexts. It becomes convenient to establish the conclusions of that particular piece of reasoning once and for all and to sum it up in something like the following form:

IF   (...conditions...)
THEN (...conclusions...)

Here, the first parenthesis is a statement of the conditions that must be fulfilled in order for the piece of reasoning to apply. The second parenthesis is a statement of the conclusions. Once the line of reasoning has been established, we can use the conclusions any time we are sure that the requisite conditions are fulfilled, without having to repeat the reasoning. These prefabricated blocks of logic are called theorems. The manner in which they are formulated and used involves the technique of *abstraction*, a basic ingredient in mathematical reasoning. The abstract nature of mathematical theorems refers to the fact that they are not statements about any particular physical entity. Rather, they are statements that may be applied to every type of physical (or conceptual) entity and in a wide variety of physical contexts (consider, for example, the statement $4+10=14$). With this in mind, the mathematician–philosopher, A. N. Whitehead, facetiously wrote that a mathematician is someone who doesn't know what he is talking about.

The construction of mathematical theorems is the primary business of the "pure" mathematician. The applied scientist often is willing to accept a mathematician's word for the correctness of the theorem. Nevertheless, it is generally a good idea to understand the principles underlying the proof of a theorem before using it, just as it is a good idea to understand the operation of a thermometer before using it. Such understanding has a certain esthetic value to the individual scientist, but of more general importance it assures that the theorem (or thermometer) is being correctly used and that the results are being correctly interpreted.

1.5. MATHEMATICAL PROOF AND PHYSICAL PROOF

The point is often made that mathematical work is either right or wrong (see, for example, Riggs, 1963, p. 4), and that if sufficient details of the work are given, one can tell from inspection whether it is done correctly. The work may be trivial or irrelevant to the question at hand, and the result may be incorrectly applied, but the mathematical work itself is either right or wrong in the following sense:

Every statement, except the starting assumptions, must follow from previous statements by strict rules of logic. The final conclusion must be a direct logical consequence of the starting assumptions.

In studying a mathematical argument, one should ask the following questions:

a. *What are the starting assumptions* (*Caution*—they are not always explicitly stated)?
b. *For each statement* (*other than assumptions*), *what previous statements is it based on, and how does the given statement follow from these previous statements*?

This prescription entails a considerable amount of care and patience, but it is the only insurance against loosing the thread of the argument.

Once a scientist is satisfied as to the correctness of the mathematical development, he or she must, before applying the result, determine that the assumptions made are indeed reasonable and applicable to the system under study.

EXERCISE 1.5. In a paper on oxygen transport in tissue, an author developed a formula for the penetration of oxygen into metabolizing cells. Among the assumptions used in the development of the formula were the following:
1. The cells are all spherical and of the same diameter.
2. Oxygen concentration on the outside of each cell ("ambient" oxygen concentration) is uniform over the whole surface of the cell. It is assumed to be the same for all cells and is held constant.

In order to test the resulting equation, would you prefer to use data on the respiratory activity of:
a. whole animals?
b. suspensions of single cells?
c. tissue slices?

Justify your answer.

1.5 MATHEMATICAL PROOF AND PHYSICAL PROOF

Unfortunately, statements that can be proved once and for all are the exceptions rather than the rule in experimental science. A statement asserting the existence of some specific physical entity is such an exception, since it could be proved by finding and exhibiting the physical entity. A statement of a general physical law, however, can only be tested but not proved. For example, the statement,

"all crows are black"

could be tested by examining a large number of crows and finding that they are all black. Finding a single non-black crow would disprove the statement, but a conclusive proof would involve examining all crows that have been or will be, and this is not possible. The best that we can do is to say that the test results are *compatible* with the hypothesis that all crows are black.[4]

The basic laws of physical science, for example of thermodynamics and quantum mechanics, as well as the generalizations of biology, share this property. They cannot be proved short of performing every one of the infinite number of experiments to whose outcome they are relevant. They are accepted as being "true" when they are found to be compatible with a large body of experimental results and seriously contradicted by none.

When the results of a mathematical argument are being used to test the validity of the starting assumptions, an additional question must be posed:

c. For each explicitly stated assumption, how was that assumption used in obtaining the conclusion?

If the assumption has not been used at all, then agreement between theoretical conclusion and experimental observation cannot be regarded as confirmation of the assumption, nor can disagreement be regarded as disproof of the assumption. In more general terms, the stringency with which the assumption has been tested depends on how sensitively the conclusions depend upon it.

A difficulty arises when the available experimental results are consistent with more than one causal hypothesis—the nonuniqueness problem again. Such instances are not uncommon and give rise to arguments in the literature over the correct interpretation of the available experimental

---

[4]This assumes that the statement to be proved is not simply a logical consequence of our definitions. If, for example, blackness were part of the definition of "crow," then a non-black crow would be a logical contradiction, and the statement, "all crows are black," would not even need to be proved.

evidence. Some of the more well-known controversies of science have centered around the nature of light, the electrical nature of the atom, origin of living organisms (the question of spontaneous generation), biological evolution, and mechanism versus vitalism in the development of individual organisms. The reader should have no difficulty in calling to mind less famous disagreements from his own area of biology.

1.6. IDEALIZATION AND SIMPLIFICATION

In discussing the definition of the concept of models, it was pointed out that the model does not mirror all properties of the system under study. This point is related to the more general need for idealizations and approximations in the study of real physical systems. Every portion of the universe is potentially affected by the rest of the universe. If we are going to study some small system, we are faced, at a minimum, with the need to either neglect or make some simplifying assumption about the interaction of that system with the rest of the universe. The question is not whether we introduce simplifying idealizations; the question is only where these idealizations should be made and how extensive they should be. Paradoxically, this may involve using physical assumptions that may be false, or even impossible to fulfill. As pointed out by Beckner (1959, Chapter 3), models built using such assumptions are useful providing there is some provision for discounting or eliminating the resulting distortion. It is this criterion that Beckner proposes as a basis for the distinction between simplification and *over*simplification. Examples from physical science are the models of the "frictionless piston" and of the "ideal gas" system. In these models, certain interactions are explicitly taken into account, and others are neglected. The effects of the explicitly considered interactions are evaluated. Disagreement between the predicted behavior and the observed behavior is then "blamed" on the neglected interactions. The disagreement is used as the basis for further refinements to the models, with the object of explicitly accounting for previously neglected interactions. Another set of examples is provided by the prey–predator interaction models of population dynamics, and is considered in Chapter 7 and Appendix A5. The earliest of these models assumed only the simplest types of interaction, and the work of refining these models continues.

Another way of simplifying a mathematical model (in addition to the use of simplifying physical assumptions) is through *mathematical approximations*. This involves replacing one mathematical expression with a simpler one that is pretty close to it under certain specified conditions.

Consider, for example, the equation,

$$y = x + x^2$$

If x is always less than 0.00001, say, then $y = x$ might be an adequate approximation. On the other hand, if x is always greater than 10,000, then the appropriate approximation would be $y = x^2$. However, for intermediate values, say x between 0.1 and 10, neither approximation would serve. The topic of mathematical approximation is further discussed in Appendix C.

1.7. MATHEMATICAL RESULTS AND BIOLOGICAL INTUITION

Before concluding this general introduction, a word of caution should be added about not neglecting one's own biological "intuition." The result of a mathematical development should be continually checked against one's intuition about what constitutes reasonable biological behavior. When such a check reveals a disagreement, then the following possibilities must be considered:

a. A mistake has been made in the formal mathematical development;
b. The starting assumptions are incorrect and/or constitute a too drastic oversimplification;
c. One's own intuition about the biological field is inadequately developed;
d. A penetrating new principle has been discovered.

Alternative d is, of course, what everyone roots for, but acute embarrassement lies in wait if the first three are not carefully checked. Agreement between intuition and the mathematics is certainly no guarantee of correctness, but disagreement should always be regarded as a danger signal. Properly used, a mathematical model is an adjunct, an aid, or a crutch to intuition; it is never a substitute.

1.8. FURTHER READING

General suggestions for further readings are largely a reflection of an author's own tastes and beliefs as to what is of benefit and interest to students. In many cases, several alternative suggestions are made; my advice is to browse through the alternatives and select what is closest to your own interest.

1.8a. On Mathematics in Biology

Bailey (1967), Bross (1972), Howland and Grabe (1972), and Smith (1968) contain general discussions on the use of mathematics in biology. Rubinow (1975) considers the subject through the development of models for several types of biological systems. The book edited by Mesarovic (1968) brings together a number of points of view on the mathematical modeling of biological systems. J. G. Miller (1973) discusses a unified system theoretic approach to the modeling of living systems.

Also of a general nature, but requiring somewhat more mathematical training, are the books by Rashevsky (1960), Roberts (1976), the volume edited by Lucas (1962), and the three-volume work edited by Rosen (1973).

Works that deal with mathematics and modeling in specific fields of biology are fairly numerous. A few examples are: Patten (1971), Pielou (1969), Smith (1974), and Watt (1966) in ecology; Milhorn (1966), Milsum (1966), and Riggs (1963, 1970) in physiology; Crow and Kimura (1970), Falconer (1960), and Mather and Jinks (1971) in genetics; Garfinkel et al. (1970) and Hemker and Hess (1972) in biochemistry. Other works are cited in connection with discussion of specific examples.

1.8b. Introductions to Mathematics

Recently, a number of works whose purpose is to teach mathematical topics to biologists have appeared. Examples are Batschelet (1975), Clow and Urquhart (1974), and Grossman and Turner (1974).

1.8c. General Works on the Modeling of Systems

Books in this category are of several different types. Auslander, Takahashi, and Rabins (1974) and Hale (1973) are useful introductions to the mathematical structures and concepts of system theory. The reader who wishes to further explore the concepts of system theory may wish to consult Klir (1972), Bertalanffy (1968), and Pattee (1973). Chapters I and II of Mesarovic, Macko, and Takahara (1970) may also be recommended.

The literature on computer simulation lays heavy stress on system modeling. Some references are given in Chapter 9.

1.8d. The Nature of Mathematics

A number of authors have tried to discuss the basic concepts of mathematics free—or as free as possible—of technical formalism. This is a diffi-

1.8 FURTHER READING

cult task. One may especially recommend Russell (1920), Newman (1956), and the three-volume work edited by Aleksandrov, Kolmogorov, and Lavrent'ev (1963).

1.8e. Philosophy of Science and Scientific Explanation

It is impossible to begin a study of explanatory modeling of systems without raising questions as to the nature of scientific explanation. For the reader interested in this topic, the books by Brody (1970), Hempel (1965), and Kuhn (1970) may be suggested as a beginning. The following deal with the problem within the specific context of biology: Beckner (1959), Elsasser (1966), Hull (1974), Rensch (1971), and the book edited by Pattee (1973), which has already been mentioned in subsection 1.8c.

CHAPTER
TWO
THE MODELING PROCESS—
AN OVERVIEW

"... and what is the use of a book,"
thought Alice,
"without pictures or conversations?"
Alice in Wonderland
Lewis Carroll

The general strategy in creating an explanatory model for a complex system is to decompose (subdivide) the system into simpler components or *subsystems*, to describe the behavior of each of the subsystems, and then to study the interactions between the subsystems. The study of the interaction pattern is frequently helped by the use of various types of diagrams. The resulting mathematical model may then be used to explain, predict, and possibly manage or control the overall behavior of the system.

In Section 2.1, this strategy is applied to the study of a simple public address system. This example is deliberately chosen to be simple enough so as to make the skeleton of the model building process clearly visible. The remainder of the chapter discusses the steps of the process in somewhat more detail. The rest of this book is essentially devoted to the further exploration of the topics raised in Chapters 1 and 2 and the application of the model-building process to biological systems.

In Appendix A, the modeling process is applied to the study of several biological systems. The examples of Appendix A have been chosen so as to be understandable (at least at an introductory level) without requiring large amounts of time in the study of specialized background material and so as to cover a broad spectrum of biological areas. I suggest that, after completing Section 2.1, the reader look through each of the examples of

2.1. A SIMPLE PUBLIC ADDRESS SYSTEM

Appendix A. At least one of the examples should be selected for more careful, continuing examination while reading through the rest of the book.

2.1. A SIMPLE PUBLIC ADDRESS SYSTEM

STEP 1. FORMULATION OF THE PROBLEM

The problem considered here concerns the build-up of a high-pitched whistling sound in a public address system. We wish to know enough about the processing of signals in the system so that we can get a clue as to how this build-up can be eliminated.

STEP 2. QUALITATIVE DESCRIPTION OF THE SYSTEM

The system consists of: a microphone, which converts sound energy into an electrical signal; an amplifier, which uses electrical energy to increase the strength of the signal from the microphone; and a loudspeaker, which converts this increased signal back to sound energy. Electrical signals and electrical energy are carried through wires, whereas the sound energy is carried through the air. The qualitative description of the system is often aided by a sketch such as the one in Figure 2.1.

STEP 3. DEFINITION OF RELEVANT COMPONENTS, SUBSYSTEMS, AND INTERACTIONS

Part of the definition of components has already been done as part of the qualitative description of the system. We are now required to make assumptions as to which of the components are important. In this example, a component is taken to be important only if it does something to the signal being processed. It is *assumed* that neither the electrical plug nor the wire that connects the physical components do anything to change the signal, and that the fate of the signal can be adequately described by

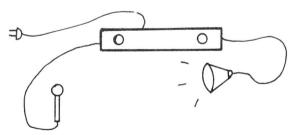

Figure 2.1. Elementary components of a public address system.

knowing what the three components (microphone, amplifier, and loudspeaker) do to the signal and how the signal passes from one component to the other. Note that the assumption that the connecting wires can be neglected would not be valid if transmission were over a very long distance. The assumptions concerning the choice of relevant components and their interconnections are shown in the component diagram, Figure 2.2. The arrows in Figure 2.2 do not represent wires; they simply indicate that some signal of interest issues from the component or block to which the tail of an arrow is attached and is received by the block to which the arrow head is attached. One speaks of the *output* and *input* of the individual components. Arrows that originate in no component of the system represent *net input* to the system. Arrows that do not terminate in any compartment of the system represent *net output* of the system. When an arrow originates in another arrow, the signal has been split. Similarly, when one arrow terminates in another, the two signals have been combined. The component diagram is a schematic representation of the structure of the system to be considered. It is not changed unless there is a need to modify that structure.

STEP 4. DEFINITION OF RELEVANT VARIABLES

The next step is to define the variables (quantities that may vary with time) whose values at any given time tell us what we want to know about the system. The problem, of course, is to know what is relevant. A useful rule of thumb is to take the variables to be the inputs and outputs of the individual components. At the same time that we define the variables, it is convenient to assign them symbols so that we may refer back to them without unduly overworking ourselves. We might, for example, construct a table such as the following.

Variable	*Symbol*
Input to system	$I(t)$
Input to microphone	$I_M(t)$
Output from microphone	$O_M(t)$
Input to amplifier	$I_A(t)$
Output from amplifier	$O_A(t)$
Input to speaker	$I_S(t)$
Output from speaker	$O_S(t)$
Output from system	$O(t)$

In this table, the (t) is meant to indicate that each of the variables may vary from one instant to the next.

2.1 A SIMPLE PUBLIC ADDRESS SYSTEM

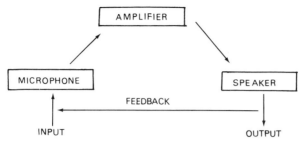

Figure 2.2. Elementary components and flow of signal for public address system.

At the time the variables are listed, it is usually desirable to specify their *dimensions*. The dimensions of a variable essentially tell what kind of physical entity the variable represents. A discussion of the dimensions for this system is deferred until Chapter 3.

STEP 5. REPRESENTATION OF THE RELATIONS BETWEEN THE VARIABLES

The way in which each variable changes is influenced by one or more of the others. The pattern of relations between the variables may be represented by a type of diagram, sometimes called a *signal–flow graph*, such as that of Figure 2.3. In this diagram, the arrows represent directions of *influence*.

The signal–flow graph may be summarized by a set of *dependency statements* of the type (see Appendix B-2).

$$X = X(A, B, C, \ldots, Z) \tag{2.1}$$

Here, X is one of the system variables, and the notation is meant to indicate that X is a function of the variables that are inside the parenthesis. Figure 2.3 translates to the following set of dependency statements:

$$I_M = I_M(I, O_S) \tag{2.2}$$

$$O_M = O_M(I_M) \tag{2.3}$$

$$I_A = I_A(O_M) \tag{2.4}$$

$$O_A = O_A(I_A) \tag{2.5}$$

$$I_S = I_S(O_A) \tag{2.6}$$

$$O_S = O_S(I_S) \tag{2.7}$$

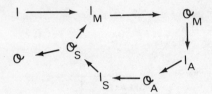

Figure 2.3. Signal–flow graph showing how one variable influences another in the public address system.

STEP 6. DESCRIPTION OF THE SUBSYSTEMS

The next step is to determine the specific form of the dependencies and to use this information to convert the component diagram of Figure 2.2 into the input–output diagram, Figure 2.4. To do this, we must examine the characteristics of the components of the system.

In the microphone, a sound signal is converted to an electrical signal. For each increment of sound energy supplied to the microphone, a part is lost because the microphone is not perfectly efficient. The remainder is converted to electrical signal. We can assume that each increment of sound energy is treated identically by the microphone, independent of how much other signal is fed in at the same time, so the total electrical signal output is proportional to the sound energy input. Letting y be output and x be input, we have $y = E_M x$, where E_M is the efficiency of the microphone and is a number between zero and one.

In Chapter 5, it is argued that the use of linear equations (proportionalities) is generally associated with increments that behave identically and

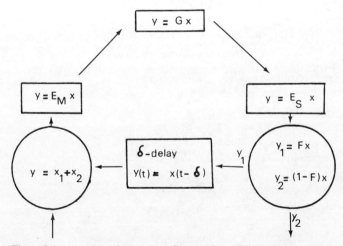

Figure 2.4. Augmented input–out diagram for public address system.

2.1 A SIMPLE PUBLIC ADDRESS SYSTEM

independently. If we make the comparable assumptions for the amplifier and speaker, that is, each increment of input is acted upon in an identical manner, independently of total size of the input, then the input–output equation for the amplifier becomes $y = Gx$, where G is the amplification, or "gain," and is generally greater than one. The input–output equation for the speaker would be $y = E_S x$, where E_S is the efficiency of the speaker.

The next component, which is not shown in Figure 2.2, but which is needed to complete the description of the system, is the atmosphere through which the sound is carried. The sound signal is of two types: the speaker output and the input to the microphone.

The speaker output must be split into the net output of the system and the part that feeds back to the microphone. To account for this, we can think of a *signal splitter*, which has two outputs for any input,

$$y_1 = Fx \tag{2.8}$$

$$y_2 = (1 - F)x \tag{2.9}$$

Notice that $F + (1 - F) = 1$, so that the signal splitter assigns a fraction F of the input to y_1 and the remaining fraction $(1 - F)$ to y_2.

The input to the microphone is the sum of the net input to the system and the feedback from the speaker. To account for this signal, we imagine a *summer*, which receives two inputs, but gives a single output,

$$y = x_1 + x_2 \tag{2.10}$$

The description is not quite complete, because we have not accounted for the time that it takes for the signal to move through the system. In general, we have to account for the time that it takes for the signal to get through each component as well as the time it takes to get from one component to another. In this case, we use the approximation that transfer of electric signal is instantaneous compared with the much longer time that it takes for sound signal to get from the speaker to the microphone. To account for this time, we need another conceptual component that shows a delay of δ time units between the splitter and the summer.

The input–output diagram of Figure 2.4 shows the input–output relations for each component, points out the interconnections between the components, and includes the three new conceptual components. This diagram comprises a mathematical description of the system. It embodies the abstract relations that are believed to be of importance in characterizing the system. For the time being, we can forget about the actual system and work with this model. In Chapter 9 we discuss the direct use of input–output diagrams as a basis for computer simulation.

STEP 7. THE MODEL EQUATIONS

For many purposes, it is convenient to express the model as a set of equations. We can do this by combining the information of the input–output diagram with the signal-flow graph of Figure 2.3 (or the dependency equations (2.2) through 2.7)) to get

$$I_M(t) = I(t) + F O_S(t-\delta) \tag{2.11}$$

$$O_M(t) = E_M I_M(t) \tag{2.12}$$

$$I_A(t) = O_M(t) \tag{2.13}$$

$$O_A(t) = G I_A(t) \tag{2.14}$$

$$I_S(t) = O_A(t) \tag{2.15}$$

$$O_S(t) = E_S I_S(t) \tag{2.16}$$

$$O(t) = (1-F) O_S(t) \tag{2.17}$$

STEP 8. STUDYING THE BEHAVIOR OF THE MODEL

Here, again, we need to make a choice. What aspects of the system behavior are of interest? In this case, we make use of the fact that the feedback has already been recognized as an important part of the problem, that is, that the signal getting to the microphone is not limited to the system input I. To see what the total input to the microphone is, we can start with equation (2.11) and proceed by successive substitutions, to get

$$I_M(t) = I(t) + F E_S G E_M I_M(t-\delta) \tag{2.18}$$

The problem is the second term on the right-hand side. Examination of the system parameters reveals that G is the easiest to adjust (a knob is usually provided) and that F may be adjusted by changing the physical arrangement of the components. Reducing F brings the $I_M \equiv I$ ideal closer to reality. So does reducing G, but this also reduces O, which may not be tolerable.

Suppose now, that we put a single short pulse into the microphone to establish $I_M(t-\delta)$ in equation (2.18) and look at I_M immediately after the

2.1 A SIMPLE PUBLIC ADDRESS SYSTEM

pulse, so that $I(t)=0$. It should clear that if

$$\frac{I_M(t)}{I_M(t-\delta)} = FE_M GE_S > 1$$

then the signal continues to build up (even though $I=0$) until the speaker cone rips or someone shuts the blasted thing off. If

$$FE_M GE_S < 1$$

then some echo effect may be produced, but the situation is at least tolerable. I_M eventually dies to zero.

This rather lengthy analysis thus tells us what most people already knew anyway: To eliminate feedback build-up in a public address system, either turn down the volume (reduce G) or move the microphone away from the speaker (reduce F). Experimental test of the conclusion demonstrates its correctness and gives us some degree of confidence in the accuracy of our conceptual understanding of the system.

In treating this example, we have been able to directly manipulate the equations to obtain a solution to the problem. Moreover, we have not been interested in the behavior of the system except in relation to the particular problem posed. Neither of these conditions is likely to hold in the study of a biological system. For very complicated systems, direct manipulation of the equations is likely to prove difficult or impossible; it becomes necessary to resort to numerical calculations on a computer. The use of computers is discussed in Chapter 9. In Chapter 7, we turn to the study of general characteristics of system behavior.

2.2. SUMMARY OF THE MODEL-BUILDING PROTOCOL

The essential thing that has been attempted with the example of Section 2.1 is to exhibit the bones of a model-building procedure. The procedure can be represented by the flow diagram of Figure 2.5. Several comments should be made immediately:

a. The process should always be considered to be iterative in that after each step, the results of all earlier steps should be reexamined and modified in response to insights gained while executing the later step.

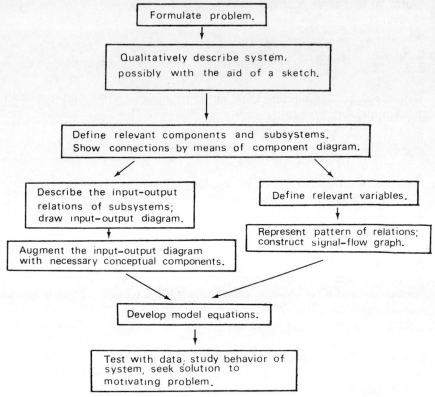

Figure 2.5. Idealization of the process of building a mathematical model.

b. All assumptions made during the process should be made as explicit as possible.
c. The whole process should be regarded as an amplification of the arrow in Figure 1.1:

$$\text{concept} \rightarrow \text{relation}$$

d. The "recipe" is intended only as a guide. Each individual must develop his or her own procedure, tailored to the particular problem at hand.

EXERCISE 2.1. Carefully examine the application of Steps 1 through 5 to at least one of the examples of Appendix A.

EXERCISE 2.2. Choose a system in your own field of biology and carry the process through Step 5.

2.2 SUMMARY OF THE MODEL-BUILDING PROTOCOL

In the following sections, we discuss some of the individual steps in the procedure.

2.3. FORMULATION OF THE PROBLEM

It is very easy to say that the first step is to define the problem, but actually doing so is usually very hard. Often one of the purposes of the model is to help in the formulation of the problem. Usually, the insight gained in the process of model building causes one to go back and reformulate the problem. Finally, it is possible to define the problem too narrowly or too rigidly, thereby becoming blind to valuable "serendipitous" insights. It is with this in mind that Chapter 1 has been opened with the "Grook" by Piet Hein.

2.4. DECOMPOSING THE SYSTEM

2.4a. General Considerations

The next series of steps leads to the decomposition (subdivision) of the system into component subsystems and a description of the network of interactions between the subsystems. How to do the decomposition is by no means straightforward. The first question is the level of description. A finer level of decomposition often leads to simpler components, but to a more complicated interaction network. For the problem considered in Section 2.1, there was no need to examine the internal circuitry of the amplifier, that is, to decompose the amplifier into its component subsystems. However, if the problem were related to a malfunction of the amplifier, then such a fine-level decomposition would be called for. On the other hand, representing the whole system as a single component, with a single overall input and a single overall output, does not permit an analysis of the problem.

Choosing a level of description, in the sense just discussed, does not complete the decomposition. Suppose, for example, the public address system had been described in terms of two physical components. Such a decomposition would have been satisfactory if the two components were a microphone and an amplifier–speaker, but the analysis would have been impossible using the components amplifier and speaker–microphone. The question of how to lump parts of the system is one that generally taxes the investigator's knowledge of the system being studied. As we have seen, the component subsystems do not necessarily correspond to physically identifiable components of the system.

The decomposition of the system and the network of interactions between the subsystems is usually displayed by means of the component diagram. The complexity of this diagramatic representation depends on how "coarse" or "fine" a description of the system is needed. Each subsystem is represented as a box. The interconnections represent inputs and outputs; the output of one box is the input to another box. Overall system inputs are inputs that do not originate as outputs from any box. Overall system outputs are outputs that do not terminate as inputs to another box.

Once the mathematical input–output relations have been added, as in Figure 2.4, we can forget what system the diagram is supposed to represent. Any system whose components have input–output relations of the same form, with the same connections, has the same mathematical description. If it suits the purpose, we may pretend that it is the flow diagram of a computer program, write the corresponding program, and thereby simulate the system on a computer.

2.4b. Compartmental Decompositions

A special type of decomposition, which is particularly convenient when it is applicable, represents the system as a set of intercommunicating *compartments*. Such models are applicable when the study concerns the flow of some entity through the system. In such models, the flow through the system is represented as being composed of transfers from one compartment to another. For example, in studying the elimination of accumulated nitrogen gas in human subjects undergoing decompression, Rescigno and Beck (1972) represent the system as the three compartments—lipoid tissue, aqueous space, and the enclosed environment.

In the component diagram for such a model each component is a compartment, and the arrows between compartments indicate transfer pathways. The system considered by Rescigno and Beck might be represented as in Figure 2.6. An example of a more complex compartmental system, shown in Figure 2.7, is adapted from a study on the distribution of DDT in the world ecosystem (Harrison et al., 1970).

The subsystems (compartments) do not always correspond to physically identifiable components. In some cases, the compartments may be taken to be states of the system under study. For example, in discussing the development of insects from eggs to adults, we may take each developmental stage as a "compartment" and discuss the transfer of individuals from

2.4 DECOMPOSING THE SYSTEM

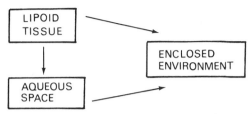

Figure 2.6. Compartments for gas washout model of Rescigno and Beck (1972).

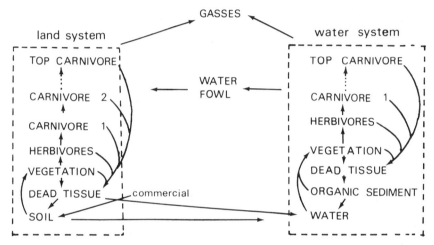

Figure 2.7. Schematic compartmental representation of flow of DDT in global ecosystem. Adapted from Harrison et al. (1970).

one compartment (instar) to another. In studying certain diseases, it may be convenient to regard each stage as a compartment and to construct a mathematical model based on the transfer between them.

The convenience of this type of decomposition is that it leads directly to a set of equations based on the simple balance relation,

(change in compartment j) = (sum of all transfers into compartment j)

$$- \text{(sum of all transfers out of compartment } j\text{)}$$
$$+ \text{(creation within compartment } j\text{)}$$
$$- \text{(destruction within compartment } j\text{)} \quad (2.19)$$

Of course, the total number of transfer pathways could be extremely large. For a system with n compartments, there could be transfers into a given compartment *from* each of the remaining $(n-1)$, transfers out *to* each of the remaining $(n-1)$, together with exchanges with the environment (the environment of the system is everything that is not considered part of the system). This leads to $2n(n-1)+2$ possible transfer routes. This is a large number, even for a three-compartment system (Figure 2.8). An essential part of the modeling process involves making decisions (assumptions) as to which to these pathways are important and which may be neglected.

The question of what constitutes creation or destruction depends upon the entity being studied. If we are looking at a population of living organisms, these might correspond to births and deaths. If we are looking at a chemical metabolite or drug, these would correspond to chemical reactions.

Equation (2.19) is a kind of conservation law, which states that anything that disappears from any compartment is either lost to the system completely (destroyed or gone to the environment), or else it reappears immediately in another compartment; anything that appears in any compartment is either new to the system (created or entered from the environment) or has come from another compartment. The utility of a compartmental decomposition depends upon the applicability of the balance equation (2.19). In turn, this may depend upon the point of view of the investigator. For example, in the prey–predator system discussed in Appendix A5, there are the three components: vegetation, prey, and predator. It might be quite useful to use equation (2.19) to discuss the transfer of some nutrient substance or even the transfer of total biomass. Equation (2.19) does not apply, however, when we study numbers in the population, since "a prey"

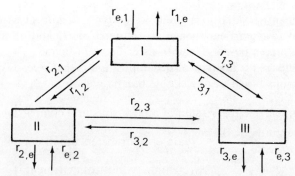

Figure 2.8. Diagram of three compartment system showing all possible transfer pathways. Notation is discussed in Chapter 6.

2.4   DECOMPOSING THE SYSTEM

does not leave the prey component and become "a predator" in the predator component.

Mathematical description of the transfers can be formulated in terms of probability of transfer or in terms of rate of transfer (see Chapter 5).

2.5. VARIABLES, PARAMETERS, AND THE STATE OF THE SYSTEM

The quantities that appear in equations (2.11) through (2.17) for the public address system may be divided into several types.

First, the *overall input* I and the *overall output* O represent the interaction of the system with its environment. In the investigation of a natural system, the overall input includes the experimental conditions and external stimuli applied to the system. The overall output includes all the observable responses. In a sense, we can think of the model as being constructed to explain the relation between the overall input and output. An investigator often modifies the system so as to obtain an output not normally yielded by the system or to render it abnormally sensitive to a particular type of input. When this is done, the investigator must carefully consider the relation between the original system he or she wishes to learn about and the modified system that is, in fact, being studied.

The *parameters* F, G, E_S, E_M, and δ are taken to be part of the structure of the system. That does not necessarily mean that they are held constant. For example, the amplifier gain may change as the transistors age or the efficiency of the speaker change because a strong input has damaged it. However, the model says nothing about the cause of these changes. In order to uniquely determine the output that results from a given input, the values of parameters (and their changes) must be separately specified. An important part of the investigation of a natural system is the estimation of the values of the parameters, and this is discussed in Chapter 8. It should be pointed out, however, that the distinction between parameters and overall inputs is not always clear. As a practical matter, quantities that show a great deal of variability are usually treated as inputs, whereas those that are more nearly constant are treated as parameters. So, for example, if we have a growing bacterial population, the rate of growth is determined by temperature, nutrient supply, acidity, and so on. The "rate of growth," which is often treated as a parameter, includes the dependence of the bacterial system upon its environment.

The last category of symbols contains the variables I_M, O_M, I_A, O_A, I_S, and O_S. These tell us what is going on inside the system at any particular time. They are not always observable, but they may be.[1] The overall output depends on the instantaneous values of the variables. They are to be taken as defining the instantaneous condition or *state* of the system and may be referred to as *state variables*. The variation of these quantities is explained by the model in the following sense: If we know the form of the equations, the value of the input, and the values of the parameters, then the model tells us how the state variables change.

A distinction may be made between inputs, outputs, parameters, and state variables on the basis of their mutual dependencies, as accounted for by the equations of the model. Inputs and parameters depend upon none of the other quantities. Outputs depend upon inputs, parameters, and state variables. State variables depend upon inputs, parameters, and each other.

Finally, we note that a number of the state variables are related to each other by simple multiplication by a parameter. In the case of the public address system, for example, complete information concerning the state of the system at any given time can be conveyed by quoting the values of two (but not *any* two) of the variables, say I_M and $\mathcal{O}_S$.

2.6. SIGNAL FLOW GRAPHS. RELATIONS BETWEEN VARIABLES

This use of the word "graph" may be unfamiliar to some readers. A general definition is that a graph is simply a collection of points, some pairs of which are connected to each other. The points are often called *nodes*. If the connections have direction—that is, are represented by arrows instead of just lines—then the graph is called a *directed graph* or *digraph*. Signal–flow graphs are thus *directed graphs* in which the nodes represent system variables.

The signal–flow graph has a particular intuitive appeal, since it represents, in pictorial form, the causal relationships between the variables that have been chosen to describe the system.

In many cases, it is convenient to attach a sign ($+$ or $-$) to the arrow to indicate the type of influence. That is, if a change in X (increase or decrease) produces a change in Y in the same direction, it is symbolized as

$$X \xrightarrow{+} Y$$

---

[1] In the case of a biological system, it is often the behavior of the internal state variables that is of primary interest. If they are directly observable, then the job is much easier. If they are not observable, then the model may be used to test hypotheses about their behavior. As mentioned before, it may be possible to modify the system so that important variables become observable.

2.6 SIGNAL FLOW GRAPHS. RELATIONS BETWEEN VARIABLES

If a change in X leads to an opposite change in Y (raising X lowers Y; lowering X increases Y), then the negative sign is used. Note that all arrows in the signal–flow graph for the public address system have positive signs.

A greater pictorial appeal is obtained if different types of arrows are used instead of just showing the sign. Riggs (1963) calls such representations symbol–arrow diagrams. In this book, we use the following symbolism: Positive arrows are shown as double arrows; negative arrows are shown as dashed arrows; arrows of uncertain sign are shown as ordinary single arrows. A number of examples are shown in Appendix A.

EXERCISE 2.3. Choose one or more examples from Appendix A and construct the appropriate symbol–arrow graphs.

EXERCISE 2.4. In Exercise 2.2, you were asked to begin construction of a model of some system from your own field of biology. Draw a symbol–arrow graph for this system.

The signal–flow graph helps in visualizing that a change in one variable leads to a change in another, which leads to a change in another, and so on. That is, the effect of changing a given variable propagates down a chain of causalities. When the chain that leads away from a given variable leads back to it again, then we have a *feedback loop*.

For example, in Figure A5.3 of Appendix A5, V, the level of vegetation, affects $\dot{P}$, the rate of change of the prey population size. This brings about a change in P, which influences $\dot{V}$, which changes V. Any alteration in V, therefore, *feeds back* upon itself. The feeding back may either magnify the original alteration (called positive feedback) or dampen the original alteration (called negative feedback).

To see if a particular variable is part of a feedback loop, we just start at that variable in the signal–flow graph and "walk" along the arrows—always in the *direction* of the arrows—from one variable to another. If we can find a walk that brings us back to where we started, then the variables that we have met along the way are all part of the same feedback loop. The same idea can be applied to the component diagram. It should be clear from looking at the signal–flow graphs of Appendix A that a given variable (or component) can be a member of several feedback loops.

The question as to whether a given feedback loop exerts a positive or negative feedback is easily determined from the symbol–arrow graph. We simply take the product of all the signs of the arrows in the loop. So, for example, in the loop of Figure A5.3, which contains V, P, $\dot{P}$, and $\dot{V}$, we have three positives and one negative, whose product is negative. The effect of the feedback loop is thus to stabilize the system against disturbances. Chapter 7 considers the topic of feedback in more detail.

EXERCISE 2.5. In the previous paragraphs, the terms "positive feedback" and "negative feedback" are introduced (more rigorous definitions of these concepts are given in Chapter 7). Examine the feedback loops of one or more systems from Appendix A, and determine whether they are positive or negative using the definitions of this section.

EXERCISE 2.6. Does the system you have begun modeling have feedback loops? If not, construct another model of another system that does. Do the loops give rise to negative or positive feedback?

2.7. DETERMINING THE FORM OF THE INPUT–OUTPUT RELATIONS

Determination of the form of the input–output relation is the main subject of Chapters 3 through 6.

The determination of the input–output relations actually involves the creation of a "submodel" for each of the subsystems. In a broad sense, the approaches used fall into the following categories:

a. Construction of a correlative submodel,
b. Mechanistic and theoretical arguments,
c. Probabilistic theoretical arguments,
d. Dimensional arguments,
e. Qualitative arguments.

This section contains a few general remarks about each of the categories.

2.7a. Direct Observation and the Construction of a Correlative Model

In this approach, we construct a correlative model of the relation between two (or more) variables, that is, of the behavior of the individual component of the system. This correlative model of the individual subsystem becomes a part of the structure of the explanatory model of the entire system. It is the usual procedure when we are interested in "explaining" the behavior of a system in terms of the behavior of its components. We

2.7 DETERMINING THE FORM OF THE INPUT–OUTPUT RELATIONS 35

are not (for the moment) interested in why the components behave as they do. We are only interested in how the behaviors of the individual components combine to produce the behavior of the system. The ecologist may not be concerned (at first) with the question of *why* the organisms exhibit the behavior they do, but only with the question of how organisms with the given kinds of behavior interact. Correlative models in this context have the same drawbacks as those discussed in Chapter 1. Principally, they offer little insight as to how the behavior of the components—and, therefore, of the system—change when the conditions are altered somewhat. For this reason, if for no other, ecologists, zoologists, and botanists must eventually concern themselves with questions of physiology; physiologists with questions of biochemistry; biochemists with fundamental chemistry. Indeed, the various disciplines of biology may be seen to divide the whole field according to a hierarchical structure. The biologist at each level of organization is concerned with the following:

i. The operation of a system at a given level of complexity (the ecosystem level, the organism level, the cell, the organelle, or an individual enzyme reaction);
ii. The behavior of the components of this system (these components are the systems of central concern at the next simpler level of organization);
iii. How individual systems that he is concerned with interact to produce a "supersystem" (the system of central concern at the next higher level of complexity).

2.7b. Mechanistic and Theoretical Arguments

Such arguments must be based on a knowledge of how the individual component of the system works, that is, a knowledge of the causative relation between the variables. Such arguments are frequently used "backwards." That is, we hypothesize a mechanism, deduce a mathematical relation between the variables, and build this relation into the mathematical model for the system. If the behavior of the model disagrees with the observed behavior of the system, then all hypotheses that have been made along the way are held suspect. As we go back to modify the model, we face the problem of ranking our hypotheses according to the degree of faith we have in them. On the other hand, if the observed behavior of the system and the behavior of the model agree, we are led to increase the degree of faith we place in all the component hypotheses.

2.7c. Probabilistic Theoretical Arguments

It often happens that we understand or are willing to hypothesize a causative relation between two variables, X and Y, but lack sufficient detail (or are unwilling to hypothesize sufficient detail) to make statements of the type, "if the input X has a certain specific value, say x_1, then the output Y always has the unique value y_1." It may be that the given input x_1 produces y_1 part of the time, y_2 part of the time, and possibly other outputs on other occasions. If one drops a rubber ball from a given height (the input) and measures the height of the bounce (the output) with sufficient accuracy, it will be found that the same input to the same ball does not always produce exactly the same output; still less, if the identical inputs are applied to distinct, but (seemingly) identical, balls. When we come to biological systems, every biologist is fully aware that no two examples of any type of biological systems are truly identical.

One solution to this problem is to say that for a given input value x_1, the average output is $y_{1,av}$, and to build the model in terms of these averages. An alternative is to list all the possible individual outputs for a given input, together with the probability of obtaining each one, and to build the model in terms of these probabilities (See Chapter 4). The first alternative is usually much easier, often sufficiently accurate, and by far the most used. Nevertheless, a general understanding of the underlying probability structure often allows a deeper insight into the operation of the system.

2.7d. Dimensional Arguments

This topic is the subject of Chapter 3. An important part of the definition of any variable is the determination of its dimensions. The equations that express relations between physical entities must be dimensionally correct if they are to make physical sense. The requirement of dimensional correctness often suggests what form the equation should have.

2.7e. Qualitative Arguments

This may be termed a last resort to a first approach. We may lack sufficient data or theory to construct a correlative or theoretical model, yet we have sufficient knowledge of a broad nature to be able to make certain statements about the relation between two variables. For example, we can say with pretty decent assurance that an animal in a scarce food situation eats everything he can find. With a superabundant food supply, he eats his fill, but there is a limit to how much he can eat. These qualitative ideas

2.7 DETERMINING THE FORM OF THE INPUT–OUTPUT RELATIONS

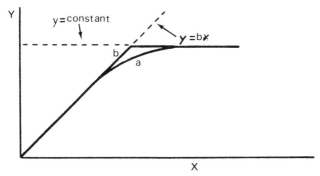

Figure 2.9. Behavior of a relation that starts in a direct proportionality and becomes: *a)* gradually saturated; *b)* abruptly saturated.

would be reflected by a curve such as (a) in Figure 2.9, where y represents food eaten and x represents food supply.

In drawing this curve, we use very general knowledge to deduce the nature of the relation at the extremes (food very scarce and food very plentiful). Next we say that as we move from one extreme to the other, the relation is unlikely to change in an abrupt way, so we draw a *smooth curve* that connects the behavior at the two extremes. You, however, may argue, "The behavior at the extremes seems reasonable, but I see no reason that the transition must be smooth rather than abrupt." With this reasoning, you might draw a curve such as (b) of Figure 2.9. A decision between the two would have to rest on appropriate observations. Indeed, such disagreements often form the basis of new experimental and field work.[2]

Qualitatively rationalized mathematical relations are often made part of the structure of the model of a complicated system. They are especially important in the early stages of a study in building a basis for further investigation.

Note that in this case the reasoning is largely geometric. Using arguments based on one's general, sometimes intuitive, knowledge about a system, a curve is drawn to represent the hypothesized relation. Then so that the relation may be considered in the mathematical framework of the model for the whole system, one finds a mathematical equation whose curve has the same qualitative behavior as the curve that has been drawn (see Section 1.2).

[2] CAUTION. When a population of individuals is being considered, the population may exhibit a smooth transition, even though the behavior of each individual changes abruptly. This happens when the individuals do not all switch at the same value of x. See, for example, Curnow (1973).

The curve that is drawn is not drawn through a specific set of data points, although all available pertinent data should be considered in its construction. Nor is the shape of the curve dictated by knowledge of a specific theoretical or "causative" relation between the variables, although any available bits and pieces of such knowledge should certainly be considered in its construction. This approach is simply a way in which one tries to codify and use the available (incomplete) information to the fullest extent.

It should be clear that the usefulness of such an approach depends not only on the artfulness of the modeller but also on his or her knowledge of the field. Moreover, since knowledge must often be liberally supplemented with hypothesis when using such an approach, it is important (though usually very difficult) to keep the two distinct. Failure to do so often leads to the publication of very scientific-sounding nonsense. Properly used, however, this general approach is very nearly an essential ingredient in the construction of models for complicated systems.

In converting the qualitatively reasoned curve to a mathematical expression, the nonuniqueness problem discussed in Chapter 1 is again with us. For any type of curve that one may draw, it may be possible to think of several kinds of expressions that have the same qualitative features, especially when the range of available data is limited. The actual choice may be made on purely intuitive grounds, pending more refined experimental observation, or it may rest on a consideration of the class of mechanisms likely to be involved.

It should be noted that certain types of mathematical expressions are connected with certain classes of mechanisms. Experience with these expressions leads one to make a mental association between the expression and the related class of mechanisms, so that the mathematical equation becomes a *physical* statement.

CHAPTER
THREE
DIMENSION AND SIMILARITY

"Can you do division?
Divide a loaf by a knife—
what's the answer to that?"
Through the Looking Glass
Lewis Carroll

When a set of mathematical equations is set forth as a description of some real world system, it must be borne in mind that the equations refer to relations between sets of physical objects or quantities. Most of us are on guard against assertions of the type, "2 oranges are equal to 2 apples." However, when the relations involve even moderately complex mathematical expressions, statements of comparable absurdity can easily slip by. The need to avoid such pitfalls is the first reason for an explicit study of dimension. Beyond this, the requirements of dimensional correctness can be valuable in the effort to understand the basic relationships that characterize a given type of system.

In the literature of physical science, little attention is given to the definition of the word "*dimension*." Instead, it is customary to begin a discussion on dimensional analysis by introducing a set of "fundamental units" or "primary units," such as mass, length, time, and temperature. These fundamental physical dimensions are discussed in Section 3.1. After reviewing the elementary rules for manipulating dimensions (Section 3.2), it is argued that the biologist needs a more general and versatile concept of dimension. The remainder of the chapter is devoted to developing such a concept and to examining its relation to the concept of *similarity*.

3.1. FUNDAMENTAL PHYSICAL DIMENSIONS

The fundamental physical dimensions are usually taken to be mass, length, time, temperature, and electrical charge.[1] They are usually assigned symbols, such as

mass	M
length	L
time	T
temperature	θ
electrical charge	Q

Each of these basic *dimensions* may be expressed in terms of a variety of *units* such as: pounds, grams, tons for mass; inches, feet, miles, meters for length; seconds, hours, years for time; degrees Farenheit, Centigrade, Kelvin for temperature. Other physical dimensions are then expressed in terms of these by considering how they arise from the "fundamental" ones. For example, the area of a rectangle may be expressed as the product of two lengths and has dimension L^2. Since the area of *any* two-dimensional region may be expressed, through the limiting process of calculus, as a sum of the areas of rectangles, the dimension of area is taken to be L^2. For an analogous reason, the dimension of volume is taken to be L^3. Velocity arises from the division of distance (length) by time, and so velocity has the dimension LT^{-1}. The rules by which such combinations are arrived at are discussed in the next section.

3.2. SIMPLE ALGEBRAIC RULES FOR DIMENSION

There are a few simple rules that govern the combination of quantities in a mathematical expression and that determine the dimension of the result.

a. Multiplication. Any two quantities may be multiplied together. The dimension (and units) of the result is the product of the dimension (and units) of the two factors. As explained previously, the dimensions of area and volume arise in this way.

b. Division. Any kind of quantity may be divided by any other kind of quantity. The dimension (and units) of the result is expressed by formally

---

[1] Discussions of electricity and magnetism often employ units of magnetic pole strength or of dielectric constant rather than units of electrical charge. (See, for example, Pankhurst, 1964.)

3.2 SIMPLE ALGEBRAIC RULES FOR DIMENSION 41

dividing the dimension of the numerator by the dimension of the denominator. The dimension of velocity, LT^{-1}, is obtained in this way.

c. Dimensionless Ratios. When a quantity is divided by another quantity of the same dimension, expressed in the same units, application of rule b gives a result with no units and no dimension. The values of such *dimensionless ratios* are often of great use in summarizing the intrinsic relations between characteristics of a system. A simple example is the ratio of wing span to total length in an airplane. The ratio summarizes an important aerodynamic characteristic that is the same for a model as it is for the working airplane and does not depend on the scale of the model.

By application of rules a and b, multiplication or division of any quantity by a dimensionless number does not alter its dimension.

d. Exponents and Logarithms. An exponent is simply a shorthand specification of an operation to be performed with a given variable. While in any given instance, the numerical value of an exponent may be based on the consideration of certain physical quantities, the exponent itself, from the very definition of what it represents, cannot be a physical entity—it cannot be associated with a dimension. However, the expression that contains the exponent (for example, the expression x^2) has a dimension that is the result of the application of the operation. If X is a length, then the dimension of X^2 would be (by rule a), L^2; the dimension of $X^{1/2}$ would be $L^{1/2}$.

Since logarithms are basically defined as exponents of a particular type (Appendix B7), it follows that logarithms also do not have dimension.

e. Conversion factors. Conversion factors are special ratios whose numerators and denominators have the same dimension, but different units. Examples are 12 in. ft^{-1} and 60 sec min^{-1}. They are constants used in the interconversion of scales associated with the same dimension.

f. Addition and Subtraction. When two quantities are added or subtracted, they must have the same dimension and the same units, which then become the dimension and the units of the result. For example, the addition 3 ft + 12 in. can be performed only after the two terms are converted to the same units by means of a conversion factor such as 12 in. ft^{-1}, to get (3 ft) (12 in. ft^{-1}) + 12 in. The addition of terms belonging to different dimensions, such as 3 ft + 10 min is, however, an absurdity that cannot be redeemed by the use of conversion factors.[2]

---

[2]Some *apparent* contradictions to this assertion are discussed in Section 3.4.

g. Equality. When two quantities are set equal to each other, their dimensions and units must match. Note that the *matching* of units does not always mean that the units are the *same*. We *might*, for example, wish to write 12 in. = 1 ft. However, if the units are not the same they had better be written explicitly, since 12 = 1 will always be nonsense.

h. Derivatives and Integrals. The dimension of a physical entity is unaffected by its magnitude. Recall that the derivative dy/dx is the limit of the ratio (change in y)/(change in x) as the two changes become small. It follows that the dimension of the derivative is given by rule (b): Dimension of y divided by the dimension of x.

An expression of the type $\int y\,dx$ is the limit (Appendix B6) of an expression of the type $\Sigma y \Delta x$. It has the dimension of y multiplied by the dimension of x.

i. Probabilities. In Chapter 4, it is seen that one way of regarding the probability of an event is as a limit of an expression of the type (number of times the event occurs)/(number of times the event *can* occur). Probabilities are, therefore, by rule b dimensionless.

A convenient way to determine the dimension of any mathematical expression is to rewrite the expression using the appropriate dimension symbols in place of the actual quantities and then to apply these rules. For the most part, the dimension symbols are treated as if they were ordinary variables and are combined by the rules of ordinary algebra—except for the constraints of rules d and f. In this way, one may arrive at the dimensions of the various physical entities found in reference books in physics and chemistry. The same procedure may be followed in determining appropriate units.

Example. Dimensions of the Variables and Parameters in the Public Address System. In the public address system of Section 2.1, the variables are of two types: electrical signals and sound signals. The electrical signals are currents, that is, an amount of electrical charge passing a particular point per unit time. Therefore, using the symbols of Section 3.1, the dimension of the variables $\mathcal{O}_M$, I_A, $\mathcal{O}_A$, and I_S becomes QT^{-1}.

The sound signals have dimension ML^2T^{-3}. To see this, we note first that sound energy is a mechanical quantity that involves the vibrational motion of molecules of air, and the resulting vibrational motion of some part of the microphone—usually a small magnet. Energy is defined as the ability to do work, and mechanical work is defined in terms of the displace-

3.2 SIMPLE ALGEBRAIC RULES FOR DIMENSION

ment of a physical object; it is the force applied times the distance moved. We start, therefore, by finding the dimensions of force, which we get from Newton's law of motion:

$$\text{force} = \text{mass} \times \text{acceleration} \tag{3.1}$$

The acceleration of a physical object is the rate of change of its velocity. Since velocity has dimension LT^{-1}, it follows from rule h that acceleration has dimension LT^{-2}. From equation (3.1), we see that force has dimension MLT^{-2}. Since energy (work) is force times distance, the dimension of energy turns out to be ML^2T^{-2}. Now, the total sound signal being received by the microphone or being emitted from the speaker is the rate at which the energy is supplied or emitted; it is energy per unit time. Therefore, the dimension of the variables I, I_M, Θ_A, and Θ is ML^2T^{-3}.

The parameters G and F are dimensionless; this follows from equations (2.11) and (2.14). The parameter E_M is current produced per unit of sound energy, or $M^{-1}L^{-2}T^2Q$. The parameter E_S has the dimension of sound produced per unit of current, or $ML^2T^{-2}Q^{-1}$. Parameter δ has the simple dimension T.

Note that the strategy followed is to use the definition of each variable to express that variable in terms of simpler quantities—signal in terms of energy and time; energy in terms of force and distance; force in terms of mass, length, and time. The dimension of the variable is then built up from the dimensions of the simple quantities, using rules a through i.

Confusion often arises between the dimension of an individual physical quantity and the dimensions of a physical system. Geometrically, a real physical object is three-dimensional, but each of the three quantities is expressed in terms of the same dimension, length. The ratio between a length in the x direction and length in the y direction is, therefore, dimensionless. Nevertheless, it is often desirable to regard each direction as a separate *dimension of the system*. To put it in a slightly different way, one may wish to treat the height and the width of a given object as two distinct properties of the *object* that may not be meaningfully added to or subtracted from each other.

In terms of the discussions of Section 2.5 on state variables and of Appendix B2.2 on ordered n-tuples as points in n-space, each state variable may be regarded as a separate dimension for the system. If the state of the system is specified as an n-tuple of values of state variables, then the system can be represented by the corresponding point plotted in the n-dimensional state space. Using this terminology, the public address system of Section 2.1 is six-dimensional, although it was seen in section 2.5 that because of the redundancies introduced by equations (2.11) through (2.17), the dimensionality could be reduced to two.

EXERCISE 3.1. Convert
a. 60 mi/hr to ft/sec;
b. 60 mi/hr to km/hr;
c. 15 lb/in.^{2} to kg/cm^2;
d. 15 lb/in.^{2} to dyn/cm^2;
e. 32 ft/sec^2 to cm/min^2.

3.3. THE CONCEPTS OF SCALE AND DIMENSION

As part of most introductory discussions on the topic of dimension, the student is told that everything he or she deals with in science must be reduced to the basic four or five "physical dimensions." It is implied that failure to do so is unscientific. I agree that it is desirable to make such a reduction whenever it is possible to do so without losing the essence of the quantities under consideration. When studying the relatively simple systems of physics, this possibility is generally open. It is often, but *not always*, open in the field of biology. For example, even though it may sometimes be desirable to express the size of an animal population in terms of total mass, doing so may lead to the loss of the essential physical content of the equations.

Although the option of reducing all quantities to fundamental physical dimensions is not invariably open to the biologist, he is *not* free from the necessity of making sure that his equations are dimensionally correct. Rather, he has the additional burden of understanding, in somewhat greater depth, the underlying nature of measurement and dimension so that he may, when necessary, *validly* coin new dimensions appropriate for the system under study.

Let us begin with a simple example. Suppose we have two length scales L and L' (say, one in inches and one in centimeters) and two mass scales W and W' (say, one in ounces and one in grams). Suppose also, that we have two physical objects, labeled a and b. If we apply each of these scales to both objects, we get eight values, which we can call L_a, L_b, L'_a, L'_b and W_a, W_b, W'_a, W'_b. The following relations obviously hold for any two objects—a and b:

$$\frac{L'_a}{L_a} = \frac{L'_b}{L_b} \tag{3.2a}$$

3.3 THE CONCEPTS OF SCALE AND DIMENSION

and

$$\frac{W'_a}{W_a} = \frac{W'_b}{W_b} \tag{3.2b}$$

However, we would *not expect* to have

$$\frac{L_a}{W_a} = \frac{L_b}{W_b},$$

except by accident. Relations (3.2) state that measurements on two scales of the same dimension always have the same ratio, no matter what object is being measured. This simple idea is the basis of the concept of dimension.

The first step is to define the concept of a *measurement scale*. If we select a physical object, application of some measurement scale gives us a unique real number. The scale may, therefore, be thought of as a function relation between a set of physical objects (the domain set) and the set of real numbers (the range set).

DEFINITION 3.1. MEASUREMENT SCALE

A measurement scale is a rule of association between a set of physical objects and the set of real numbers. The association involves a physical operation (measurement). Given any element of the domain set, the rule specifies a unique value in the range set.

It should be clear that L, L', W, and W' meet this definition. The next step is to determine when two scales are of the same dimension, that is, when they measure the same physical characteristic. The answer is provided by relations (3.2), which can be stated in a general way.

DEFINITION 3.2. BELONGING TO SAME DIMENSION

Any two measurement scales, say S and S', are said to *belong to the same dimension* if they have the same domain set and if

$$\frac{S'_a}{S_a} = \frac{S'_b}{S_b} \tag{3.3}$$

where S_a, S'_a, S_b, S'_b are the measured values for *any* pair of objects a and b from the domain set.

To say that a relation such as (3.3) holds for any pair of objects is to say that the ratio S'/S is the same for *every* object from the domain set.

We may, of course, have a whole collection of measurement scales, any pair of which obey relation (3.3). This collection or set of measurement scales is what we call a dimension.

DEFINITION 3.3. DIMENSION

A *dimension* is a set of measurement scales, specified by the characteristic that for every pair of scales S and S' that belong to the set, the ratio of a measurement on S to the same measurement on S' is constant and independent of what is being measured.

The two foregoing definitions are slightly different ways of saying the same thing.[3,4]

As a simple example, suppose we take as our domain set the set of all physical objects that can be said to contain another physical object, for example, the set of all "containers" (boxes, vessels, etc.). Suppose we have a measuring process that consists of counting the number of dozens of apples in the container, and we call the result A_D. Another process counts the number of gross of apples with result A_G. Similar processes measure oranges with results $\mathcal{O}_D$ and $\mathcal{O}_G$. Then for any two containers (elements of the domain set) a and b, it is true that

$$\frac{A_{D_a}}{A_{G_a}} = \frac{A_{D_b}}{A_{G_b}} \qquad \text{and} \qquad \frac{\mathcal{O}_{D_a}}{\mathcal{O}_{G_a}} = \frac{\mathcal{O}_{D_b}}{\mathcal{O}_{G_b}}$$

---

[3]At first glance, it would seem that °F and °C do not meet the requirements of definition 3.2 for belonging to the same dimension. Yet, as indicated in Section 3.2, temperature is normally taken as a fundamental dimension. The contradiction is only apparent, however, for the following reason. Each of the two measurement scales (°F and °C) expresses the difference between the temperature of the measured object and that of some fixed reference object. The seeming contradiction arises because they use different reference objects: freezing water for °C and freezing saturated salt solution for °F. If readings are adjusted to the same fixed reference (for example freezing water or "absolute zero"), then we have

$$\frac{\text{degrees on Centigrade scale}}{\text{degrees on Farenheit scale}} = \frac{5}{9}$$

[4]Definitions 3.2 and 3.3 may sometimes be regarded as too narrow. That is, if measurements on two scales have a one-to-one relation (Appendix B2), then it may be argued that they really measure the same physical property and should be regarded as belonging to the same dimension. However, only if the simple proportionality of definitions 3.2 and 3.3 is used, do we have available the convenience of the rules of dimensional analysis, such as those of Section 3.3.

3.3 THE CONCEPTS OF SCALE AND DIMENSION

However, except in special cases, it is not true that

$$\frac{A_{D_a}}{\mathcal{O}_{D_a}} = \frac{A_{D_b}}{\mathcal{O}_{D_b}}$$

Thus A_D and A_G belong to one dimension and $\mathcal{O}_D$ and $\mathcal{O}_G$ to another. That is, "apples" is one dimension in which "containers" may be measured and "oranges" is another.

As part of the study of any system or type of system, it is usual to choose[5] a minimum number of fundamental dimensions in terms of which all other dimensions may be expressed through the rules of Section 3.3. Measurement scales (units) are chosen for the fundamental dimensions and units for other dimensions follow from the rules of Section 3.3. These rules prohibit an addition such as

$$2 \text{ apples} + 4 \text{ oranges}$$

However, in such a case, we could measure their mass and add the two masses. Or, we could convert both apples and oranges to the dimension "fruit" and say 2 fruit + 4 fruit = 6 fruit. *Note that in either case, a certain amount of information is lost.*

The ratio S'/S is the *conversion factor* between scales S' and S. Its units are units of S' "divided" by the units of S. A measurement in one scale can always be converted to a measurement in another scale of the same dimension by multiplication with the appropriate conversion factor.

Example. A chemical system might be characterized by the amounts or concentrations of the constituent chemical species. Each chemical substance would then be considered as a separate dimension of the system.

Example. An ecological system might be characterized by the amounts or densities of the constituent biological species. Each type of organism would be considered as a separate dimension of the system.

EXERCISE 3.2. Cite one or more examples from your own field of biology of pairs of scales that belong to the same dimension. Discuss the dimensions and units of the scale conversion factors.

[5]The choice is more often made implicitly than explicitly.

EXERCISE 3.3. Cite one or more examples from your own field of biology of pairs of scales that do *not* belong to the same dimension.

EXERCISE 3.4. Be sure that you understand the dimensions of the variables and parameters in the examples of Appendix A.

3.4. APPARENT CONVERSION OF ONE DIMENSION TO ANOTHER

If all apples were exactly of the same mass, any scale that measured mass of apples and any scale that measured number of apples would belong to the same dimension. Now suppose we consider a collection of apples, all of which are *approximately* of the same mass. For the special purpose of *that* collection, one might propose the use of an approximate conversion factor between mass and number count. However, the definitions of the previous section specify that measurements on any two scales belonging to the same dimension must always be exactly in constant fixed ratio. The scales associated with the basic physical dimensions meet this requirement *by the way they are defined*; they are *defined* as being proportional to each other. This is a fundamentally different situation from the one cited for the interconversion of number and mass on a domain set artificially limited so that the two characteristics are approximately interconvertible. Nevertheless, the use of approximate special purpose conversion factors and the treatment of two different physical characteristics as being *for the purpose at hand* in the same dimension, is a convenience that is often used. It is true that the practice is a frequent source of error. When an equation is derived for use with a limited set of objects, using a special purpose conversion factor, it may be misapplied to a wider set of objects. The best way to avoid such errors is through clear and explicit statements about the domain of validity of the conversions and of the resulting equations. Some examples of the treatment of different dimensions as being the same for special purposes are:

a. The amount of some chemical substance in solution may be expressed in terms of moles (which indicates number of molecules), mass, or volume. The conversion factors are specific for each substance.
b. When a truckload of apples is delivered to a packaging plant, a random sample from the truckload may be both counted and weighed. An approximate conversion factor may then be computed that applies only

3.4 APPARENT CONVERSION OF ONE DIMENSION TO ANOTHER

to that truckload. The number of apples may then be "counted" by weighing the load.

c. In modeling the flow of energy through an ecosystem, one must account for the storage and release of chemical energy as the atoms rearrange into different molecular forms. In such a context, a given mass of carbohydrate may be taken as being equivalent to a certain amount of energy. This is a special purpose equivalence, which says, in effect, that 1 g of carbohydrate is a package that contains a certain amount of energy that can be made available for certain purposes under certain conditions. For these purposes we treat the number of packages, for example, the number of grams, as being a direct measure of the amount of available energy. Thus amount of energy and number of grams are treated as being of the same dimension.

A dimension is a label attached to a number that gives the information "number of what." The rules discussed in this chapter are primarily to insure that the labels on various numbers are compatible with each other, and with the way the numbers are to be combined. Incorrect labeling normally leads to wrong results. Incomplete labeling can lead to a substantial loss of information.

3.5. DIMENSIONAL HOMOGENEITY

An equation is termed *dimensionally homogeneous* if its validity does not depend on the specific units used. That is, so long as the same units are used every time a given dimension appears, the relationship is independent of the particular choice of units. For example, relation (3.1) is unaffected by using pounds instead of grams or feet instead of centimeters. It is usually not desirable to have the description of a system depend on the units that happen to be chosen for measurement. As a rule, therefore, the equations of a model should be constructed to be dimensionally homogeneous.

When such equations are used in purely symbolic form, there is no need to specify the actual units used. When numbers are put in, *then* the units must be specified, and one must be careful that every time a given dimension is used, the same units are specified.

If dimensional evaluation of the expressions on the two sides of an equation shows them to disagree in dimension, then the equation cannot be dimensionally homogeneous. The cause of this condition must be determined before the equation can be used. The cause may be that a pertinent conversion factor has been neglected, that the dimensions of proportionality constants have been improperly treated, or that the whole equation is utter nonsense.

Lack of proper attention to dimensional homogeneity can often lead to a great deal of confusion. An excellent example, which, we now look at in some detail, is the confusion between the terms *mass* and *weight*.[6]

Because of the way weight is defined, it has the dimension of a force, which, as we have seen, has dimension MLT^{-2}. A unit of force that is often used is the *dyne*, which is defined as the amount of force to accelerate 1 g by 1 cm sec^{-2},

$$\text{dyne} = \text{g cm sec}^{-2}$$

Weight is defined as the force produced on some given mass by the acceleration of a gravitational field g,

$$W = mg \tag{3.4}$$

so it has dimensions MLT^{-2}. Now, if we take the figure $g = 980.665$ cm sec^{-2} as being the "average" gravitational pull on the earth's surface, we have

$$W = m \times 980.665$$

This gives the weight in dynes. If we forget the dimensions on the "constant" 980.665, it looks as if we are writing, force $= MLT^{-2} = M$. The confusion is generally compounded by converting dynes to a new unit of force, called *gram-weights*. We let

$$1 \text{ gram-weight} = 980.665 \text{ dyn}$$

In these new units, we get

$$W(\text{dyn}) \times \frac{1 \text{ gram-weight}}{980.665 \text{ dyn}} = W(\text{gram-weights})$$

$$= (980.665 \; m) \frac{1}{980.665}$$

$$= m(\text{grams})$$

---

[6]Perhaps in an age of space travel, there is less confusion on this point than there used to be.

3.5 DIMENSIONAL HOMOGENEITY

or

$$W = m \qquad (3.5)$$

Without writing the dimension of the number 980.665 (LT^{-2}), it looks as if we are asserting that a force is equal to a mass, $MLT^{-2} = M$. The distinction (lost by neglecting to keep track of dimensions) is that mass is an intrinsic property of an object, whereas weight is the force produced by that property under the acceleration of a given gravitational pull. The dimensionally inhomogeneous equation (3.5) might find a limited usefulness, if it is agreed to restrict all accelerations to that of a "standard gravitational field."

EXERCISE 3.5. Select one or more equations, more or less at random, from a recent issue of a journal in your field, and analyse each for dimensional correctness.

EXERCISE 3.6. In Chapter 2, you were asked to begin construction of a model of a system in your own field of biology. What are the dimensions of the variables for that system?

3.6. SIMILARITY

The validity of a dimensionally homogeneous equation does not depend upon the choice of fundamental units, or measurement scales. Furthermore, dimensionally homogeneous equations do not depend upon the size of the system being described (often called the scale of the system). In equation (3.1), the fundamental dimensions are mass, length, and time. In asserting the truth of the relation, we also assert that the validity does not depend on how *much* mass or length or time is associated with the system. Rather, it is asserted that all systems of a certain type are *similar* in that they all behave according to such a relation.

A dimensionally homogeneous equation thus expresses an intrinsic characteristic of the system, which continues to hold when any of its fundamental dimensions are "scaled" up or down. In this sense, a system of one size can serve as a physical model for a system of another size. The idea of scaling a system up or down is first introduced in an intuitive way but is made more precise in subsection 3.6c.

3.6a. Scale-Independent Ratios

Frequently, we wish to point out that two comparable systems are similar in that a certain characteristic has the same value for both systems, even though the two systems are of different sizes. For example:

i. Two solutions might be considered chemically similar if the concentrations of all components are the same; that is, if the ratio

$$\frac{\text{weight of component } A}{\text{volume of solution}}$$

is the same in both solutions for all components. If the volume of solution is scaled up (doubled, say), the solution remains chemically unchanged if the weights of all components are scaled up by the same factor (that is, also doubled).

ii. In studying energy flows in an ecosystem, one way in which two systems might be considered comparable is in having the same amount of incident sunlight per unit area; the ratio (incident light/area) would be the same for both systems.

iii. Two food materials might be compared on the basis of energy content per unit weight. The two materials might be judged similar in this respect if for any given samples of the two materials, the ratio (metabolizable energy content/sample weight) were the same.

iv. Two chemostats (Appendix A2) might be comparable if the ratio,

$$\frac{\text{number of new organisms per unit time}}{\text{volume of the growth chamber}}$$

were the same in each.

v. Two thermostated water baths (Appendix A3) might be compared on the basis of the ratio of the capacity of the heating element to the surface area over which heat is lost.

In each case, a relevant characteristic has been defined in terms of a ratio. Two different systems may be compared on the basis of this ratio, without regard to the actual "sizes" of the systems, and are considered similar with respect to this characteristic if they both have the same value.

Such a ratio is sometimes called scale-independent, or scale-invariant. That is, it is taken to express a characteristic of the system that is independent of the scale upon which the system is built or measured.

3.6b. Dimensionless Ratios

If the numerator and denominator of a ratio are of the same dimension (and expressed in the same units), then the ratio is dimensionless.

Dimensionless ratios are a fortiori scale-independent. More than that, if a relevant characteristic of a system can be expressed as a dimensionless ratio, then it is not necessary to keep track of which units or which measurement scales were actually used. Two different systems may be compared without concern for the "size" of either system and without concern as to the units used for either system.

i. Example. The shape of a circle is characterized by the ratio,

$$\pi = \frac{\text{circumference}}{\text{diameter}}$$

Numerator and denominator both have the dimension L, so the ratio is dimensionless. The value of this ratio is an intrinsic property of any circle and does not depend upon either the size of the circle or the particular units being used to express length.

ii. Example. In toxicity studies, small animals are frequently used as "model systems." Results are often expressed in terms of the ratio (weight of drug or toxin/weight of animal) required to produce a specified effect. While the result certainly depends upon the specific metabolic characteristics of the animal, it is used (at least provisionally) as being independent of the size of the particular animal.

iii. Example. Suppose we wish to characterize the adequacy of the food supply for some animal population. The first impulse might be to use the ratio of food available to number of animals. Dimensions of such a ratio might be taken as (mass) (animals)$^{-1}$. If the system were scaled up or scaled down, adequacy of the food supply might be taken as unchanged if this ratio remained the same. Such a ratio is useless, however, if we wish to compare the food supply in two systems that contain different animals with different food requirements. In such a case, we need to introduce a new quantity, food requirement per animal, which also has the dimension (mass) (animal)$^{-1}$. Dividing one by the other gives the dimensionless quantity,

$$\frac{\text{food available per animal}}{\text{food required per animal}}$$

Such a quantity can be used to characterize the relative abundance of food supply as an intrinsic property of the system under study and can serve as a basis for comparing systems of different sizes and with different species. Similar arguments can of course be made in characterizing the adequacy of soil nutrient supply for plants.

iv. A "Counter Example": Danger of Incomplete Labeling. Suppose we have two solutions: solution I is made by mixing 10 ml of ethyl alcohol with 100 ml of water; solution II is made by mixing 10 ml of acetone with 100 ml of water. The concentrations of both solutions, expressed as

$$C_V = \frac{\text{volume of solute}}{\text{volume of solvent}}$$

is 0.10 and is apparently dimensionless. However, 10 ml of ethanol weighs approximately 9.2 g, while 10 ml of acetone weighs approximately 7.9 g. Thus measuring concentrations by the "dimensionless ratio,"

$$C_W = \frac{\text{weight of solute}}{\text{weight of solvent}}$$

gives a value of 0.092 for solution I and 0.079 for solution II. This example is meant to illustrate two points:

a. The value of a dimensionless ratio does not change when measurement scales are changed only as long as the change is to another scale in the same dimension;
b. The relation between two dimensionless ratios (concentrations of alcohol and of acetone) may depend upon their history, that is, upon the dimensions used for the original measurements.

3.6c. Geometric Similarity; Corresponding Points and Scaling

The idea of geometric similarity is first introduced in the context of a particular example. Later in this subsection, the concept is generalized, so that it may be used in a wider variety of contexts. The example is the estimation of the surface area of an animal from its weight.

Estimation of surface area is important for instance in the consideration of energy balance and heat loss (Appendix A3). Since it is difficult to measure, there have been various attempts to relate the surface area of an

3.6 SIMILARITY

animal to its volume. Although volume is a very difficult thing to measure directly, we can usually convert from volume to weight (in the sense of Section 3.4) by assuming that the density (mass per unit volume) and the gravitational field are constant (see for example, Cowgill and Drabkin, 1927; Riggs, 1963). Now, with a relatively easy method for measuring volume, we can try to develop a relation between surface area A and volume V, which does not depend on the size of an individual animal. The ratio A/V is not appropriate, since it has dimension L, and, therefore, cannot be independent of size. On the other hand, the ratio,

$$k = \frac{A}{V^{2/3}} \tag{3.6}$$

is dimensionless. For convenience, we might refer to this ratio as the *shape factor*. The use of such a ratio can be easily illustrated for some simple shapes. For example, if all animals were spherical, we would have

$$A = 4\pi r^2$$

$$V = \frac{4}{3}\pi r^3$$

In calculating the ratio k, the radius cancels between the numerator and denominator, giving

$$k = \frac{4\pi}{(4\pi/3)^{2/3}}$$

$$= 4.83598$$

Shape factors for some other simple shapes are shown in Table 3.1.

Table 3.1 Shape Factors for Simple Geometric Shapes

Shape	Surface area	Volume	$k = A/V^{2/3}$
Sphere with radius r	$4\pi r^2$	$\frac{4}{3}\pi r^2$	$4.83598\ldots$
Cube with edge length l	$6l^2$	l^3	6.000
Cylinder of radius r length l	$2\pi(r^2 + rl)$	$\pi r^2 l$	$\dfrac{2\pi(r^2 + rl)}{(\pi r^2 l)^{2/3}}$
Cylinder with $l = c \cdot r$	$2\pi(1+c)r^2$	$c\pi r^3$	$2\pi^{1/3} \dfrac{1+c}{c^{2/3}}$

The point of this discussion (which is established more rigorously in the following paragraphs) is that objects of the same shape all have the same value for the shape factor, $k = A/V^{2/3}$. If the value of k is determined for one animal, from measurement of V and A (at cost of great effort), then the value of A for an animal of the *same shape* may be obtained from a volume measurement, using the relation,

$$A = kV^{2/3} \tag{3.7}$$

Of course, relation (3.7) holds with a constant value of k, only for objects with the "same shape"; Table 3.1 makes it clear that the shape factor changes as the type of shape is changed. Furthermore, the shape factor for cylinders depends upon the relation between radius and length and is not a constant unless this relation is held constant. It should be apparent that a more precise concept of similarity of shape is required.

In order to develop such a concept, we suppose that we have two objects, A and B. If it happens that a *point-to-point correspondence* can be established between them, then in some sense they are models of each other. Assuming that this is the case, we can designate the position of a particular point p in object A by the ordered triple (x_p^a, y_p^a, z_p^a). The position of the corresponding point p' in object B can be designated by the triple $x_{p'}^b, y_{p'}^b, z_{p'}^b$).

Next suppose that it is possible to orient the two objects relative to the Cartesian coordinate system so that every point p in A has the following relations with its corresponding point p' in B:

$$x_{p'}^b = K_x x_p^a \tag{3.8}$$

$$y_{p'}^b = K_y y_p^a \tag{3.9}$$

$$z_{p'}^b = K_z z_p^a \tag{3.10}$$

where the values of the proportionality constants K_x, K_y, and K_z don't depend on which pair of corresponding points p and p' have been chosen. If we can orient the two objects so that equations (3.8) through (3.10) hold, then the two objects are *scaled models* of each other. The constants K_x, K_y, and K_z are the *scale factors* for the lengths in the X-, Y-, and Z-directions, respectively.

As an example in two dimensions, imagine a picture drawn on a rubber sheet, and suppose that the sheet is stretched to double length in the X-direction. Then we would have $K_x = 2$, but $K_y = 1$. An example in two dimensions for which $K_x = K_y$ would be the relation between a photo-

3.6 SIMILARITY

graphic slide and the image projected on a projection screen. In general, two objects are considered to be *geometrically similar*, if the three scale factors are identical; that is, if $K_x = K_y = K_z$.

Now, suppose we consider two points p and q of object A, and their corresponding points p' and q' of a geometrically similar object B. The distance between p and q is given by the Pythagorean formula of elementary geometry,

$$d(p,q) = \left[(x_p - x_q)^2 + (y_p - y_q)^2 + (z_p - z_q)^2 \right]^{1/2} \quad (3.11)$$

Since all coordinates of object B are multiplied by the same factor K, we have

$$d(p',q') = \left[(x_{p'} - x_{q'})^2 + (y_{p'} - y_{q'})^2 + (z_{p'} - z_{q'})^2 \right]^{1/2}$$
$$= \left[(Kx_p - Kx_q)^2 + (Ky_p - Ky_q)^2 + (Kz_p - Kz_q)^2 \right]^{1/2}$$
$$= Kd(p,q) \quad (3.12)$$

Since p and q can be any two points, it follows that if two objects are geometrically similar, *all point-to-point distances* have been scaled by the same factor. This requirement excludes the third entry of Table 3.1; cylinders are not geometrically similar in this sense unless the ratio of radius to length is fixed.

Finally, if all point-to-point distances have been scaled by the same factor K, then every small area element on the surface is scaled by the factor K^2. Recall that the total area can be built up as the limit of an "infinite number of infinitely small" volume elements (see Appendix B5), so the total surface area is scaled by the factor K^2. Analogously, every small volume element, and, therefore, the total volume, is scaled by the factor K^3. The shape factor, $k = A/V^{2/3}$, is, therefore, the same for two objects that are geometrically similar.

In the practical application of equation (3.7), it must be realized that no group of animals has precisely the same shape factor k. However, if the animals are sufficiently alike, one may determine an average shape factor. As discussed in Appendix D, it is appropriate to use a logarithmic rather than an ordinary arithmetic average.

Considerations of geometric similarity become important in taxonomic and evolutionary studies (see, for example, Lewis, 1976), although the literature does not always take full advantage of the formalism discussed in this subsection.

EXERCISE 3.7. For the system you are modeling, what dimensionless ratios can you find? What scale-invariant ratios characterize meaningful properties of the system?

3.6d. More General Concepts of Scaling, Similarity, and Corresponding Points

In the previous subsection, the ideas of scaling and of geometric similarity were introduced by means of corresponding points in three-dimensional physical space. The same ideas can be introduced in a more general context through the definition of corresponding points in the n-dimensional state space of the system. Since each point of the state space represents a possible state of the system, corresponding points of state spaces represent corresponding states of two systems. In analogy with the discussion of the previous subsection, two systems may be taken as scaled models of each other if it is possible to establish a one-to-one correspondence between the states of the systems, and if the following property holds. If we let X be any state variable (dimension of the state space), s denote any arbitrary state of system A and s' the corresponding state of system B, $x_s^{(a)}$ be the value of X for system A, and $x_{s'}^{(b)}$ be the value of X for the corresponding state of system B, then

$$x_{s'}^{(b)} = K_x x_s^{(a)} \tag{3.13}$$

where the same scale factor for the X variable holds for every pair of corresponding states.

Two systems may be said to be *similar* if the scale factors for all (or all relevant) state variables are the same. When two systems are similar in this sense, it follows that all relevant dimensionless ratios have the same values for the two systems.

In a *dynamic system* (dynamic systems are more explicitly discussed in Chapter 5), the state of the system continues to change with time, so that the point representing the system moves through the state space. For such systems, the idea of corresponding points can be applied to the $(n+1)$-dimensional space in which time is added as the $(n+1)$st dimension. When the behavior of two similar systems is such that they move through corresponding states at corresponding times, the two systems are said to be *dynamically similar*.

To take a simple example, when a small animal is used as a physical model for a larger animal, we can establish corresponding times in the life cycles (relative, for example, to the events of birth, weaning, puberty, etc.).

3.6 SIMILARITY

If W_t^a is the weight of animal A at time t and $W_{t'}^b$ is the weight of animal B at corresponding time t', then the two animals could be said to have similar growth patterns—that is, to be dynamically similar with respect to the single state variable weight—if the ratio,

$$K_w = \frac{W_t^a}{W_{t'}^b}$$

is the same for every pair of corresponding times t and t'.

In experimental biology, it is commonplace to extrapolate an observation made on one type of system to another, which is in some sense "similar," that is, to use one biological system as a model for another. In this section, we have tried to indicate some of the factors that must be taken into account in assessing the validity of such extrapolations. One must also realize that similarities between biological systems almost always have to be considered to be of an approximate, rather than of an exact, nature.

3.7. SCALE LIMITATIONS

If a biological system could be scaled up or down without affecting the relevant dimensionless ratios (or, more generally, the scale-invariant ratios), then there would be neither an upper nor a lower limit on the size that any type of system could have. In general, however, this *cannot* be done. The reasons involve the unalterability of certain physical constants, which cannot be arbitrarily scaled up and down, and the relation between the system and its environment.

As an example, one may cite the finite load-bearing capacity of biological structures. Ability to withstand direct compression can be expressed as a certain weight (force) per unit area. Since the load per unit area increases directly with height, it follows that there is a physical limit for the height of trees, as well as for land animals, quite apart from other considerations, such as bending and twisting stresses (see, for example, McMahon, 1973).

As another example of a limitation on size, consider a single living cell that depends for its metabolic activity upon the diffusion of nutrient materials from the environment and upon the diffusion of waste materials out into the environment. Suppose the cell is spherical. As the radius of the sphere gets larger, the volume of the cell, and so the demand for nutrient material, increases as r^3. But the rate of supply is limited by the surface area in contact with the environment, which increases only as r^2. Therefore, a given type of cell, with a given type of metabolism and shape, cannot become arbitrarily large.

On the other hand, the establishment of the relationships that characterize the system often requires a minimum number of components, each with its own minimum size. For example, a cell, in order to have the properties that would characterize it as a living cell, would have to store a minimum amount of information in molecular form (the size of atoms and molecules puts a lower limit here) and possesses a minimum number of enzyme molecules. On this basis, one may conjecture as to the minimum size of a living cell (Morowitz, 1967). Analogously, one might conjecture as to the minimum complexity needed to insure stability of an ecological system. Such conjectures have been the subject of vigorous debate because of their implications for public conservation policy (see, for example, Diamond and others, 1976).

Size limitations of the type discussed in this section are not unique to biological systems. They are a concern of the engineer as well. For example, the behavior of a small model of a harbor is affected by the surface tension of water. This is a force whose magnitude is constant. Even though it is negligible in comparison to other forces in the full size harbor, it might easily be comparable to the forces encountered in a small model. On the other end of the scale, the height of man-made structures is clearly subject to the kind of limitations discussed for trees and land animals.

3.8. FURTHER READING

Chapter II of Riggs (1963) contains an elementary discussion of dimensions and units. More detailed treatments may be found in Pankhurst (1964) and in Langhaar (1951), which is more theoretical and which goes into more detail on the subjects of scaling and similarity. The book by Ellis (1966) discusses basic concepts of measurements, dimensions, and units.[7]

---

[7]The reader who consults Ellis will find that he argues at length against the inclusion of counting measures, such as were used in Section 3.4, as valid dimensional scales. In my view, these objections can be met by considering that because of the quantum nature of physical reality as described by theoretical physics, *all* physical measurements are reducible to counting measures.

CHAPTER FOUR
PROBABILITY MODELS

"But, surely
God does not throw dice...."
Albert Einstein

In Chapter 2, the principle focus is on the structure of a model for a system, that is, the definition of important components and variables and their pattern of interrelationships. Chapter 3 continues with a consideration of the qualitative nature of the variables as reflected by their dimensions. It was seen that a systematic consideration of dimension can lead to greater understanding of the pattern of relationships between the variables. In this chapter, we begin to look at the explicit formulation of the model equations.

Most often, when we describe the behavior of some system, the description is in terms either of totals or of averages. For example, we speak of deer born in a given year, or the total acetyl methyl carbinol produced by all the *Escherichia coli* cells in a test tube. When we speak in terms of individual members of a population, we normally divide the total by the estimated population size and refer to the average behavior of the individuals. We do not pretend that each individual exhibits this identical behavior. If we speak of the birth rate of some type of animal, we don't mean that each member of the population has that number of offspring in each year. Rather, we generally are indicating a figure that is an average of the whole population over many years. The average is intended to indicate a number around which the individual values are expected to bunch. It is understood that the behavior of the individuals is to some extent random, but that out of this underlying randomness there arises, in some way, the reasonably predictable regularity of the aggregate system.

Mathematical models are generally formulated on the basis of the observed aggregate regularities, without regard to the randomness of the individual behavior. In this book, however, the purpose is to explore the relation between the physical concepts that relate to individual behavior and the model that describes the whole system. We, therefore, begin the study of the formulation of model equations with a direct consideration of the random character of the underlying events. At the same time, it is well to point out that descriptions of many types of systems *require* explicit consideration of the underlying probabilities. Examples may be found in almost every field of biology.

Since many students of biology have never had formal training in probability theory, the first part of this chapter is devoted to a study of the elementary principles of probability. In order to provide a specific example for this discussion, Section 4.2 introduces one of the famous genetic experiments of Mendel. Readers who are familiar with the elementary principles of probability may wish to read Section 4.2 and then skip to Section 4.11.

The references listed in subsection 1.8b have chapters with more complete introductions to probability theory. For students with a background in probability, Karlin (1975) might be suggested as an introduction to theory of stochastic (probabilistic) processes. Goel and Richter-Dyn (1974) discuss stochastic models in a number of specific areas of biology.

4.1. PROBABILITIES AND WEIGHTED AVERAGES

The most commonly used type of average is the arithmetic average or *arithmetic mean*.[1] It is simply the sum of a whole bunch of numbers divided by the total number of numbers,

$$\overline{X}_{\text{arith.}} = \frac{1}{n} \sum_{i=1}^{n} x_i \qquad (4.1)$$

However, if there is only a very small number of *distinct* values of X, then the computations can be simplified by taking a *weighted* average of these distinct values. Suppose, for example, that we make measurements on the number of red-wing blackbirds caught in a special trap in one particular region in a single day. In a sequence of 100 such measurements, we might find, say, 4 with no catch, 20 with one bird, 35 with two, 25 with three, 10

---

[1] Other types of averages are discussed in Appendix D.

4.1 PROBABILITIES AND WEIGHTED AVERAGES

with four, 6 with five, and no traps with more than five birds caught in a single day. If x_i is the number of birds in the ith measurement, then the average number of birds per trap per day would be $(1/100)\sum_{i=1}^{100} x_i$. However, if we notice that we have only six distinct values (that is, the set of observed outcomes is $\{0, 1, 2, 3, 4, 5\}$), then we can let z_i stand for the ith one of these outcomes. If we let n_i be the number of times outcome z_i is observed, then the average number of birds per trap per day could be written as the weighted average of the z_i

$$\overline{X}_{\text{arith.}} = \overline{Z}_{\text{wt.av.}} = \frac{\sum_{i=1}^{6} n_i z_i}{\sum_{i=1}^{6} n_i} \quad (4.2)$$

where $\sum_{i=1}^{6} n_i =$ total number of measurements. Expression (4.2) is a *weighted* average of the values of z_i. Each z_i is multiplied by an appropriate weight that reflects its contribution to the total, that is, the frequency with which z_i has occurred. The *relative frequency* is

$$\text{relative frequency of } z_i = \frac{n_i}{\sum_{i=1}^{6} n_i} \quad (4.3)$$

Now, imagine that the trapping experiment is repeated many, many times under identical conditions in identical regions (or in the same region, assuming that the characteristics of the region remain unchanged). We expect that at the start, while the total number of trappings (which is the denominator of equation (4.3)) is small, the relative frequencies given by equation (4.3) will keep changing and fluctuating as additional results are obtained. However, as the total number of trappings becomes larger, we would expect that the fluctuations would become less and less, and that the relative frequencies would settle toward some limiting value. It is this limit that is called the probability of the outcome z_i. It is a measure of our expectation of getting that particular outcome. The mathematical characteristics of this type of limit and its relation to the type of limit discussed in Appendix B3 is part of the study of advanced probability theory. For the present purpose, it is sufficient to say that a mathematical probability may be interpreted as the relative frequency of occurrence over the long run, where the term "long run" implies a run of infinite length. Relative frequencies measured over a short (that is, finite) run may be taken as *estimates* of the probabilities. Thus expression (4.3) gives an estimate of the probability of getting the value z_i in some randomly chosen trap on some randomly chosen day. Indeed, the purpose of performing the trapping experiment might be to obtain these probability estimates.

If we let $RF(z_i)$ be the relative frequency of the possible outcome z_i, then combining equations (4.1) through (4.3),

$$\overline{X}_{\text{arith.}} = \overline{Z}_{\text{wt.av.}} = \sum_i z_i \cdot RF(z_i) \tag{4.4}$$

In keeping with this discussion, we might expect that as new experiments are added, this value fluctuates, but stabilizes as the number of measurements gets larger and larger. For a large enough number of measurements, the $RF(z_i)$, therefore, approximate the probabilities of the individual outcomes, denoted by $P(z_i)$. When the $RF(z_i)$ in (4.4) are replaced by the $P(z_i)$, the result is called the *expectation* of the variable Z.

DEFINITION 4.1.   EXPECTATION

If an experiment has r possible outcomes, denoted by the numbers $z_1, z_2, \ldots, z_r$, with probabilities $P(z_1), P(z_2), \ldots, P(z_r)$, then the *expectation* or *expected value* of the outcome is

$$\mathcal{E}(Z) = \sum_{i=1}^{r} z_i P(z_i) \tag{4.5}$$

4.2. EXAMPLE FROM MENDELIAN GENETICS; DEFINITION OF THE PROBLEM

The interpretation of the classical experiments of Mendel is one of the earliest uses of probability in modern science. We look at a particular one of these (see I. H. Herskowitz, 1962, Chapter 2), which involved the cross-fertilization of two varieties of garden pea. One of these varieties produced only pink flowers, and the other produced only white flowers. The F_1 cross between them produced only pink flowers. The F_2 produced plants with pink and plants with white flowers. If the F_2 pink-flowered plants were crossed among themselves, they produced both pink- and white-flowered plants (Figure 4.1). Seven hundred five (75.9%) were pink-flowered, and 224 (24.1%) were white-flowered. Mendel's explanation of these results may be summed up in the following list of assumptions:

4.2 MENDELIAN GENETICS; THE PROBLEM

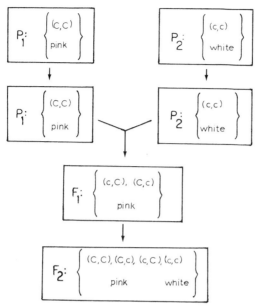

Figure 4.1. Hypothesized sequence of genotypes and related phenotypes in succeeding generations.

a. When two plants mate, each plant contributes half the genetic material that determines the characteristics of the offspring. It is not that each parent contributes the genetic material that determines half the traits of the offspring, but that each trait is determined by two genes, one from each parent. For convenience, we denote the gene leading to pink flowers by capital C, and the gene leading to white flowers by small c.

b. Since each parent is also an offspring, it also carries, by assumption a, two genes for each trait—one from each parent. The question is, which of these two genes is contributed to a given one of its own offspring? Assumption b is that this is "pure chance"; either one is equally likely.

c. The dominance assumption is that as long as either of the two genes carried is C, then the plant produces pink flowers. It produces white flowers only if both genes are c.

If these three assumptions are accepted then it immediately follows that if any group of plants has been inbred for many generations *without* incidence of pink flowers then gene C is lacking from the population. Similarly, inbreeding without incidence of white flowers indicates the absence of gene c.

At this point, we have an observed result and a hypothetical explanation that involves (assumption b) an appeal to the laws of chance. However, the hypothesis is in terms of simple events that relate to the transmission of the genetic material, whereas the observations are in terms of complex events compounded from the simple ones and relate to the phenotypic characteristics of collections of plants. We, therefore, need a formalism that allows us to express these compounded events in terms of the simple events. We need to formalize the concept of probability as well as the concepts of simple and compound events in order to see how the probabilities of different types of events relate to each other.

We begin by setting down the ideas of assumptions a through c in a mathematical shorthand. The genetic material of interest is represented by an ordered pair. The first term is the allele that comes from the female parent, and the second term is the allele from the male parent. Since one of the parents carries only gene C, its gene pair may be written (C,C). Similarly, the other parent may be written (c,c). If we cross female (c,c) with male (C,C), all pairs produced are (c,C). A male (c,c) crossed with female (C, C) produces only (C,c). According to assumption c, all have pink flowers. Thus far the prediction is in accord with observation.

If equal numbers of reciprocal crosses are made, then the F_1 generation has equal numbers of (c,C) and (C,c). Four different matings are, therefore, possible within the F_1 population—male and female (c,C) and (C,c). Taking a particular one of these, male (c,C) and female (c,C), we have a set of four possible F_2 genotypes, shown in Figure 4.1. According to assumption c, the first three would produce pink flowers, whereas only the last one shown would produce white flowers. This same set of four F_2 outcomes can be seen to result from any of the four F_1 types of mating combinations.

4.3. THE SAMPLE SPACE OR SET OF OUTCOMES

As a prelude to a discussion of probabilities in connection with any experiment or group of experiments, we need to list all the possible outcomes to that experiment or group of experiments. The *set* of outcomes should exhaust all possibilities, and so is said to be *exhaustive*. The listing of the outcomes in the set should be done in such a way that two of them can never happen simultaneously. The outcomes then are *mutually exclusive*. To say, then, that the set should be exhaustive and the outcomes mutually exclusive is nothing but a fancy way of saying that every time the

4.3 THE SAMPLE SPACE OR SET OF OUTCOMES

experiment is made or the measurement taken, we get one and only one element (outcome) in the set. Such a set is often called a *sample space*. The last box in Figure 4.1 shows two different possible sample spaces for a single F_2 pea plant. That is, whatever its genotype, it must be one and only one of the ordered pairs in the F_2 genotype set. On the other hand, whatever its phenotype, it must be one and only one of the elements of the F_2 phenotype set.

It is convenient if the individual outcomes in the sample space are all equally likely. This is convenient because it allows us to say that if there are N elements in the set, each of these outcomes has one chance in N of being the result of any given experiment. If the experiment were repeated a great many times, we should expect that each of the outcomes would occur approximately $(1/N)$th of the time, so that each outcome has a probability of $1/N$. Such a sample space is called an *equal likelihood* sample space.

For example, if we flip a "fair" coin, the sample space might be {heads, tails}. It has two elements, and they are assumed to be equally likely, and each would be said to have probability $1/2$. If we roll a single "fair" die, the outcome set would have six elements, they would be assumed to be equally likely and so each would have probability $1/6$. In the garden pea example, it seems intuitively clear, though we have not yet proved it from the model assumptions, that the outcomes of the F_2-genotype sample space are equally likely, with probability $1/4$. However, in the case of the bird trapping example referred to earlier, the set $\{0, 1, 2, 3, 4, 5\}$ seemed to be exhaustive and the outcomes are mutually exclusive, but they are certainly *not* equally likely. Similarly, in the F_2-phenotype sample space of the garden pea example, the outcomes are exhaustive and mutually exclusive, but they are not equally likely. When the outcomes are not equally likely, their relative likelihoods must be explicitly specified.

Specification of the elements of the sample space and of their relative likelihoods (whether equal or otherwise) is the starting point for the formulation of a probabilistic model. The mathematical formalism cannot tell us what should be in the sample space or what the relative likelihoods should be. That information must come from a knowledge of the system. The probabilities of the outcomes in the sample space may be estimated from experimental determination of relative frequencies, as in the bird-trapping experiment of Section 4.1, or they may be made part of the assumptions of the model, in which case they will be accepted or rejected depending upon the performance of the model. For the F_2 sample spaces of the garden pea example, the probabilities are deduced from assumptions that relate to still simpler events: the transmission of individual alleles from parent to progeny.

4.4. EVENTS AND THEIR PROBABILITIES

In a formal sense, the term "event" means a *subset of the sample space*. An *elementary event* is a subset containing only a single element of the sample space. Subsets that contain more than one outcome are called *compound events*. An event is said to have *occurred* if any of the outcomes in it has occurred. For example, suppose we want the event that a fair die has landed even. This event can be represented by the subset $\{2,4,6\}$. If any one of the outcomes in this subset occurs, then the event *even* has occurred. Another example would be to ask about the event that a given F_2 plant is homozygous. The event *homozygous* is a subset of the F_2-genotype sample space,

$$\text{homozygous} = \{(C,C), (c,c)\}$$

If the sample space S is an equal likelihood set, the probability of any event $A \subset S$ may be computed by the following:

$$P(A) = \frac{\text{number of outcomes in } A}{\text{total number of outcomes in } S} \tag{4.6}$$

Equation (4.6) may be taken as a definition of the probability of any event $A \subset S$ providing that the elements of the sample space are equally likely. If we accept that the outcomes of the F_2-genotype sample space are equally likely, the probability of homozygous is $2/4 = 0.5$. For any elementary event (contains only a single outcome), equation (4.6) gives simply $1/N$.

Applying the rules of set algebra to events, we see that any compound event can be built up as a union of smaller events. For example,

$$\text{homozygous} = \{(C,C), (c,c)\}$$
$$= \{(C,C)\} \cup \{(c,c)\}$$

In general, if we have any two events A and B, their union $(A \cup B)$ is a new event, just as the union of two sets is a new set. If an outcome is in either A or B (or both), then it is in the new event $(A \cup B)$. Therefore, the new event $(A \cup B)$ is said to have occurred, if either of A or B have occurred.

When the two events A and B are disjoint (disjoint subsets of S), we can

4.4 EVENTS AND THEIR PROBABILITIES

get the probability of the union in the following way:

$$P(A \cup B) = \frac{\text{number of outcomes in } A \cup B}{\text{number of outcomes in } S}$$

$$= \frac{\text{number of outcomes in } A + \text{number of outcomes in } B}{\text{number of outcomes in } S}$$

$$= P(A) + P(B) \qquad (4.7)$$

Note that if A and B are not disjoint, then the number of elements in $A \cup B$ is not the sum of the number of elements in the two separate sets. In such a case, any elements that are in both sets would be counted twice. We would get the right answer, then, if we subtract them out once,

$$P(A \cup B) = \frac{\text{number of outcomes in } A \cup B}{\text{number of outcomes in } S}$$

$$= \frac{\text{number of outcomes in } A + \text{number of outcomes in } B - \text{number of outcomes in both } A \text{ and } B}{\text{number of outcomes in } S}$$

$$= P(A) + P(B) - P(A \cap B) \qquad (4.8)$$

In equation (4.8), the event $(A \cap B)$ contains outcomes that are in both A and B. It is, therefore, said to have occurred if *both* A and B have occurred. For example, the event that a particular plant is homozygous with pink flowers can occur only if both *homozygous* and *pink flowers* occur,

$$\text{homozygous pink} = \text{homozygous} \cap \text{pink flowers}$$

Equation (4.7) can be easily extended to unions of any (finite) numbers of events. Any time an event V can be represented as a union of a collection of pairwise disjoint events—that is, any time we can write

$$V = \bigcup_{i=1}^{r} A_i; \quad \text{with } A_k \cap A_j = \emptyset \text{ for } k = 1, \ldots, r$$
$$j = 1, \ldots, r$$
$$j \neq k$$

then it follows that

$$P(V) = \sum_{i=1}^{r} P(A_i) \qquad (4.9)$$

The event V is said to occur if any of the A_i occur. Although equations (4.7) and (4.9) have been rationalized with reference to an equal likelihood sample space, they may be used to get the probability of any union of disjoint events (Section 4.10).

The event *homozygous* has been seen, from equation (4.6), to have probability of 0.5. We can also calculate it from equation (4.7) (or equation (4.8)). Since

$$\text{homozygous} = \{(C,C)\} \cup \{(c,c)\}$$

and since these two component events are disjoint, we have

$$P(\text{homozygous}) = P\{(C,C)\} + P\{(c,c)\}$$
$$= 0.25 + 0.25$$
$$= 0.5$$

Using both equations (4.6) and (4.7) and the F_2-genotype sample space, the reader should verify that

$$P(\text{pink flowers}) = 0.75$$
$$P(\text{white flowers}) = 0.25$$

The probability of event (homozygous $\cup$ pink flowers) *cannot* be computed from (4.7) since the component events are not disjoint. As an exercise, the reader should compute this probability from equation (4.8).

4.5. SIMPLE COUNTING RULES

Equation (4.6) should make it obvious that the calculation of probabilities often involves the counting of numbers of possible outcomes. When the sample spaces are small, this can be done simply and explicitly. As the sample space gets larger, however, explicit counting becomes not only tedious but virtually impossible. Simple rules for calculating the number of

4.5 SIMPLE COUNTING RULES

various types of combinations and rearrangements are, therefore, an important part of the study of probabilities.

The counting rules that we look at in this section can be most easily *derived* by considering the number of multiples in the Cartesian product between sets. However, the rules can be most easily *applied*, as will be seen, by reformulating the counting problem as a sampling problem.

4.5a. Number of Multiples in a Cartesian Product

The first question we need to answer is: If we have a set A with n_a elements a_i ($i = 1, 2, 3, \ldots, n_a$) and another set B with n_b elements b_j ($j = 1, 2, 3, \ldots, n_b$), how many distinct ordered pairs (a_i, b_j) can be formed. Remember that two ordered pairs, say (a_i, b_j) and (a_i', b_j'), are equal to each other only if $a_i = a_i'$ and also $b_j = b_j'$. Otherwise, they are distinct. To get the answer, we first pick a single element of A, say a_1. For each separate element of B, we get a new ordered pair. This gives us n_b ordered pairs: (a_1, b_1), (a_1, b_2),…,(a_1, b_{n_b}). This is repeated with a_2, giving the n_b ordered pairs (a_2, b_1), (a_2, b_2),…,(a_2, b_{n_b}). Again, this is repeated with each of the n_a elements of A, giving n_a groups of n_b ordered pairs. The result is that the number of pairs in the Cartesian product $A \times B$ is $n_a n_b$.

As an example, suppose someone is traveling from Chicago to France by way of New York. She can travel to New York by means of automobile, train, bus, or airplane. From New York she can travel to France by boat or airplane. The total number of different combinations is obviously eight. We can let the set A be the set of ways of getting from Chicago to New York. Set B would then be the set of ways of getting from New York to France. Each individual combination would then be an ordered pair of the type (a_i, b_j). The total number of such ordered pairs is (4)(2) = 8.

By a similar type of argument, we find that if there are three sets A, B, and C, with n_a elements, n_b elements, and n_c elements, respectively, the total number of distinct ordered triples is $n_a n_b n_c$.

More generally, if there are r different sets, we get the following theorem.

Theorem 4.1. Multiples in an r-fold Cartesian Product

IF there is a total of r sets, the first with n_1 elements, the second with n_2 elements, and so on,

THEN the total number of distinct ordered r-tuples that can be formed with the first element from set one, the second element from set 2, and so on (i.e., the number of r-tuples in the r-fold Cartesian product of all the sets) is $n_1 n_2 n_3 \ldots n_r$.

The result of this theorem is sometimes put in terms of the number of ways a combination of tasks can be performed. If there are r tasks to perform, and they must be performed in order (task 1 first, followed by task 2, etc.), and there are W_1 ways of performing the first task (e.g., four ways of getting from Chicago to New York), W_2 ways of performing the second, and so on, then the number of ways of performing the combination of tasks (in order) is $W_1 W_2 \ldots W_r$.

4.5b. Counting as a Sampling Problem

We start off by considering a group, or set, of N elements that are clearly and distinctly labeled, $a_1, a_2, \ldots, a_N$. From them, we pick a sample of r elements. The first job is to count how many distinct samples are possible. The count depends on the two questions: When do we call two samples distinct? and How do we do the sampling?

There are two usual ways of answering the first question, which correspond, respectively, to equivalence of ordered r-tuples and equivalence of sets. The first method of counting ("with order") counts two samples as the same if they have the same elements in the *same order*. Otherwise, they are distinct. The second method ("without order") counts two samples as being distinct only if at least one element appears in one sample that does not appear in the other. There is a general relation between these two methods of counting. It involves the number of ways a sample can be reordered. If a given sample of r elements can be reordered W_{ro} ways, each of these reorderings would be counted as distinct if we are counting *with* order, but all would be identical counting without order. If W_{ord} and W_{unord} are the respective number of ways of choosing a sample, the two would be related by

$$W_{ord} = (W_{unord})(W_{ro}) \qquad (4.10)$$

With regard to the method of sampling, there are two different stylized methods. One is *with replacement*. In this method, any element of the population may show up any number of times (up to r times) in a sample of size r. The usual analogy is to consider that the elements are individually labeled balls in a box. We pick one ball, write down the selection, and throw it back before picking the next. The other procedure is to say that in any sample of size r, no element may appear more than once. This method is termed sampling *without replacement*. This method is analogous to choosing a ball from the box and *not* replacing it before picking the next one. Sampling with replacement allows the choosing of a sample of any

4.5 SIMPLE COUNTING RULES

size, whereas sampling without replacement obviously limits the sample size to the size of the population.

We first consider sampling *with* replacement. In this case, the first element is chosen from a set of size N. The second element of the sample is again chosen from a set of size N and so on. It is as if we were to form the r-fold Cartesian product in which all the r factors are identical, and so have the same number of elements. The total number of distinct (with order) samples possible would be the same as the number of ordered r-tuples in the Cartesian product. From Theorem 4.1, this is N^r.

On the other hand, if the sampling is done without replacement, then we again have an r-fold Cartesian product, but each succeeding factor is reduced in size by one. That is, we have N ways of choosing the first element, $N-1$ ways of choosing the second element, and so on. The number of ways of choosing, say, the kth element is the number that are left before the kth element is chosen. This is N minus the number that have already been chosen, that is, $N-(k-1)$, or $N-k+1$. Multiplying all the r factors together, the number of possible different samples (with order) is

$$N(N-1)(N-2)\cdots(N-r+1) \tag{4.11}$$

Note once again that sampling with replacement, r may be any positive whole number at all. If the population size is 20 (say, amino acids), we may form sequences (proteins) of any length. The number of distinct protein molecules that may be formed with length of 100 amino acids, for example, would then be 20^{100}. If each amino acid were chosen completely at random (each amino acid equally likely in each of the 100 positions), then each of these 20^{100} would be equally likely, and each particular one would have probability of 20^{-100}—a very small number. A particular gene that causes the formation of a protein of this size must impart to the synthesizing machinery the information required to repeatedly select the same sequence out of the total 20^{100} possible sequences.

EXERCISE 4.1. Given four different types of nucleotides, how many different sequences can be formed of length two? of length three? Why does this show that if the codes of all amino acids are nonoverlapping and of the same length, then that length must be three?

In the case of the garden pea experiment, the population that is being sampled is the set of outcomes for a single plant. From this population of outcomes, we select a sample of 929 outcomes. Since each outcome may appear any number of times (up to 929), it falls under the heading of sampling with replacement. Using the single plant outcome set {pink, white}, we have $N=2$ and $r=929$, so the total number of possible sequences would be 2^{929}.

The question of how many of these 2^{929} sequences correspond to the observed result of 705 pink and 224 white can also be answered in terms of sampling, but the sampling must be formulated in a different way. We imagine that we have a sequence of 929 blank spaces. We select 705 of these blank spaces, and to those 705 we assign the label "pink." Formulating the sampling in this way, it becomes sampling *without* replacement. The population that is being sampled is the population of 929 blank spaces, and each can be chosen at most once. However, we don't care at all about the order in which the spaces are chosen (i.e., we don't care whether position 4 was chosen first and position 10 second or vice versa), so before the answer is calculated we need to look at sampling not counting order. This is done by determining W_{ro} (number of ways an ordered sample can be reordered) and then using equation (4.10).

W_{ro} can be gotten from expression (4.11) by just letting $r=N$; that is, the number of rearrangements of any N things is

$$N(N-1)(N-2)\cdots(3)(2)(1).$$

This number should be recognized as $N!$ (N-factorial). Similarly, the number of possible rearrangements of the r elements in the sample is $r!$. Notice that this result has been obtained with the assumption that no element appears more than once; it applies only to the case of sampling without replacement. Determination of the number of rearrangements when sampling *with* replacement involves considerably more work and is not needed for the task at hand. The interested reader might refer to Feller (1957) for more discussion.

Now we have just about finished the development of the simple counting rules we need. To summarize:

1. Counting order and sampling with replacement, the number of samples is N^r.
2. Counting order and sampling without replacement, the number of possible samples is $N(N-1)\cdots(N-r+1)$. In order to write this number in a more convenient form, we notice that multiplication by $(N-r)!$ produces $N!$. The number of samples can, therefore, be written $N!/(N-r)!$.
3. Using equation (4.10), the number of samples not counting order and

4.5 SIMPLE COUNTING RULES

sampling without replacement is

$$\frac{N!}{r!(N-r)!}$$

This particular type of number arises often in various branches of mathematics, and it is assigned the symbol

$$\binom{N}{r} = \frac{N!}{r!(N-r)!}$$

It is called the "binomial coefficient N over r."

For easy reference, these results are tabulated in Table 4.1.

Table 4.1. Number of Distinct Samples of Size r That May Be Drawn from Population of Size N

Sampling	Counting	
	With Order	Without Order
With replacement	N^r	
Without replacement	$\dfrac{N!}{(N-r)!}$	$\binom{N}{r} = \dfrac{N!}{r!(N-r)!}$

Note that sampling without replacement and without order simply divides the original set of N elements into two subsets (order does not count), one of size r and the other of size $N-r$. This is sometimes spoken of as dividing the original population into subpopulations. That is, the number of ways in which a population of size N can be divided into two subpopulations, one of size $n_1(=r)$ and the other of size $n_2(=N-r)$ is $N!/n_1!n_2!$, where $n_1 + n_2 = N$.

A comparable result may be developed for dividing a population of N things into any number of subpopulations, say r of them. To do so, we can proceed in a stepwise manner. We first select the first subpopulation of n_1 elements. This can be done in $N!/[n_1!(N-n_1)!]$ ways. The remaining population has size $(N-n_1)$. Out of it we can select subpopulation 2 of size n_2. This can be done $(N-n_1)!/[n_2!(N-n_1-n_2)!]$ ways. The total number of ways in which the two subpopulations can be selected (one following the other) is, therefore,

$$\frac{N!}{n_1!(N-n_1)!} \cdot \frac{(N-n_1)!}{n_2!(N-n_1-n_2)!} = \frac{N!}{n_1!n_2!(N-n_1-n_2)!}$$

The third subpopulation is obtained from the remaining $(N-n_1-n_2)$, and this may be done in $(N-n_1-n_2)!/[n_3!(N-n_1-n_2-n_3)!]$ ways. In this way, we continue to select the subpopulations. The $(r-1)$st subpopulation is selected from the remaining $N-n_1-n_2-\cdots-n_{r-2}$ elements and can be chosen in

$$\frac{(N-n_1-n_2-\cdots-n_{r-2})}{n_{r-1}!(N-n_1-n_2-\cdots-n_{r-1})!}$$

ways. Finally, the last subpopulation is the remaining n_r elements. Multiplying together the number of ways of getting the successive subpopulations, we get the result,

The number of ways of dividing a population of N things into r subpopulations so that the first has n_1 elements, the second has n_2 elements and so on, is

$$\frac{N!}{n_1!(N-n_1)!} \cdot \frac{(N-n_1)!}{n_2!(N-n_1-n_2)!} \cdot \frac{(N-n_1-n_2)!}{n_3!(N-n_1-n_2-n_3)!} \cdots \frac{\left(N-\sum_{i=1}^{r-2} n_i\right)!}{n_{r-1}!n_r!} \cdot 1$$

$$= \frac{N!}{n_1!n_2!n_3!\cdots n_{r-1}!n_r!} \qquad (4.12)$$

For example, if we have 1000 F_2 plants, the number of sequences that have equal numbers of (c,c), (C,c), (c,C), and (C,C) would be $1000!/(250!\,250!\,250!\,250!)$.

In the garden pea example, we are interested in the question of how many sequences of 929 plants could give rise to the observed result of 705 pinks and 224 whites. As a sampling problem, this would be the question of how many ways a population of 929 can be divided into two subpopulations of sizes 705 and 224. The answer is $929!/705!\,224!$.

In the next section, we see that each of these ways has probability (according to the Mendelian hypotheses) of $(0.75)^{705}(0.25)^{224}$.

4.5 SIMPLE COUNTING RULES

EXERCISE 4.2. Suppose you have a chamber with 90 flies in it and drop in 10 additional flies. You wait for a few minutes and select a random sample of 10 flies from the chamber. What is the probability that you get back exactly the same 10 flies you added? (This problem is adapted from an experiment on genetics of flying ability in flies.)

EXERCISE 4.3. Suppose we have a collection of N amino acid molecules with n_1 of type one, n_2 of type two, up to n_{20} of type 20, with $\Sigma n_i = N$. Derive an expression for the number of different types of protein that can be made with these N molecules (use all N for each type).

4.6. COMBINATIONS OF INDEPENDENT EVENTS

In the development of equation (4.8), we had to be concerned with the combination (intersection) of two events A and B—that is, with the event that both A and B occur. In this section, we see how to express the probabilities of such combined events in terms of the probabilities of the individual events, providing that the individual events do not influence each other.

Let us start with a look at the F_2-genotype sample space. Each outcome is really the combination of two still more elementary events: getting a particular allele from the F_1 female parent together with a particular allele from the F_1 male parent. For each of the two equally likely (assumption b) possibilities for the first of these, there are two equally likely possibilities for the second, giving four equally likely possibilities, so the F_2-genotype sample space is an "equal likelihood" set; all its outcomes are equally likely, with probability 1/4.

To carry it a step further, suppose we want to get the probability that if we select two F_2 plants at random, they both have pink flowers. Looking at the first plant in terms of the F_2-genotype sample space, there are three equally likely ways for it to be pink. For each of these, there are three equally likely ways for the second to be pink, so that there are nine equally likely ways of realizing the combined event {pink first} ∩ {pink second}. Reasoning in this same way, we find a total of 16 possible combinations, so the probability of two pinks is 9/16. Similar reasoning shows the probability of the sequence {pink first, white second} to be 3/16.

In each of these cases, the number of ways of getting the combined event is the product of the number of ways of getting the individual events, so that the probability of the combined event is the product of the probabilities of the individual events. This is true, because *the occurrence or nonoccurrence of the second event is unaffected by the first*. The probability that the second plant has white flowers is unaffected by the color of the first plant and vice versa. The two are said to be *stochastically (probabilistically) independent*.

The ability to express the joint probability of two events as the product of their individual probabilities is so closely related to the intuitive idea of the independence of the two events, that it is taken as a mathematical definition of the concept of independence.

DEFINITION 4.2. INDEPENDENCE OF EVENTS

Two events A and B are said to be stochastically *independent* when

$$P(A \cap B) = P(A)P(B)$$

Using this definition and the assumption that the two plants receive their genes independently (really part of the model), we can write

$$P(\{\text{pink first}\} \cap \{\text{white second}\}) = P\{\text{pink first}\}P\{\text{white second}\}$$

$$= \frac{3}{4} \cdot \frac{1}{4}$$

$$= \frac{3}{16}$$

On the other hand, we might ask about the probability of getting one pink and one white in two trials, not caring which came first. We note that the two-plant outcome set contains two elements that correspond to the event {one pink, one white}. They are mutually exclusive, so we can simply add their probabilities,

$$P\{\text{one pink, one white}\} = P\{(\text{pink, white})\} + P\{(\text{white, pink})\}$$

$$= \frac{3}{16} + \frac{3}{16}$$

$$= \frac{3}{8}$$

When we look at a sequence of three plants, we need to ask about the joint probabilities of three elementary one-plant events. That is, we need probabilities of the type $A \cap B \cap C$. If we write it as $(A \cap B) \cap C$, and

4.6 COMBINATIONS OF INDEPENDENT EVENTS

assume that the events $(A \cap B)$ and C are independent, then we have

$$P((A \cap B) \cap C) = P(A \cap B)P(C)$$

If A and B are also independent, then

$$P(A \cap B \cap C) = P(A)\,P(B)\,P(C)$$

CAUTION. Even if A and B are independent, B and C are independent, and A and C are independent, it does not always follow that $(A \cap B)$ and C are independent. That is, $P(C)$ might be affected by a *combination* of other events.

In the garden pea example we are discussing, the outcome relating to any plant is taken to be independent of what happens to any other plant or to any combination of other plants. With this assumption we can look at a typical sequence of 929 outcomes. We pick one of these sequences that has 705 pinks and 224 whites—say the first 705 pink and the following 224 white. Extending the argument for three plants, we get the result that the probability of this exact sequence is a product whose first 705 factors are all 0.75 and whose last 224 factors are 0.25. That is, we get $0.75^{705}\, 0.25^{224}$.

But this is only one of the 929-tuples that give 705 pinks and 224 whites. The total probability of the distribution must be gotten by finding all such 929-tuples (there are $929!/705!\,224!$ of them), getting each of their probabilities and then, since they are mutually exclusive, adding them up. We are helped by one nice circumstance: Every one of the 929 plants has the same set of probabilities, $P\{\text{pink}\} = 0.75$ and $P\{\text{white}\} = 0.25$. Therefore, the probability of every 929-tuple with 705 pinks and 224 whites has 705 factors of 0.75 and 224 factors of 0.25. Only the arrangement differs:

$$P\{\textit{any given}\text{ sequence with 705 pinks and 224 whites}\} = 0.75^{705}\, 0.25^{224}$$

Therefore,

$$P\{705 \text{ pinks and } 224 \text{ whites}\} = \frac{929!}{705!\,224!} 0.75^{705}\, 0.25^{224}$$

$$\cong 0.02525$$

This is a very small number; it is tempting to conclude that the observed value is rather improbable, if the hypotheses are correct. In a strict sense, this conclusion is valid; if the experiment were to be repeated, it would be most unlikely to yield exactly the same result. However, it would be very likely to yield a result that is very close. There are 930 possible outcomes to this experiment: 0 pink; 1 pink;...; 929 pink. When an experiment has a very large number of outcomes, it almost always follows that the likelihood of getting any specific one—even the predicted outcome—is very small. In this case, the predicted result (see Section 4.8) is 697 pinks. Applying the same reasoning process as before, the probability of this exact result is found to be $(929!/697!232!).75^{697}.25^{232} \cong 0.03055$. On the basis of the model, therefore, the observed value is 82.7% as likely as the predicted value. Most people would conclude that the observed result is consistent with the hypothesized model.

Applying the same reasoning process in more general terms, we have the following theorem.

Theorem 4.2. The Binomial Distribution

IF
1. An experiment has two possible outcomes called, say, a_1 and a_2.
2. We let $P\{a_1\} = p$, so that $P\{a_2\} = (1-p)$.
3. The experiment is repeated N times and each outcome is independent of any combination of other outcomes.

THEN The probability of getting $\{a_1\}$ exactly n times and $\{a_2\}$ exactly $N-n$ times is

$$P[a_1(n \text{ times}), a_2(N-n \text{ times})] = \frac{N!}{n!(N-n)!} p^n (1-p)^{N-n} \quad (4.13)$$

In the title for Theorem 4.2, the term *distribution* is used in the following sense: A *probability distribution* is a pattern according to which every possible relevant event may be assigned a probability. Various types of probability distributions pertain to different classes of phenomena. Part of the study of probability theory involves the explicit study of more commonly encountered probability distributions.

To recapitulate, equation (4.13) was arrived at by the following sequence of steps:

a. When N experiments are performed, there may be many sequences that give $\{a_1\}$ n times and $\{a_2\}$ $N-n$ times. In the case of the garden pea

4.6 COMBINATIONS OF INDEPENDENT EVENTS

experiment, $\{a_1\} = \{\text{pink flowers}\}$ and $\{a_2\} = \{\text{white flowers}\}$; $n = 705$ and $N - n = 224$. Many different sequences of 929 plants could give 705 pinks and 224 whites.

b. Any one of the sequences has probability $p^n(1-p)^{N-n}$. For the garden pea experiment, the assumption is that $p = 0.75$, so any one sequence with 705 pinks and 224 whites would have probability $0.75^{705} 0.25^{224}$.

c. The sequences are mutually exclusive, so their probabilities simply add. Since all the sequences with $\{a_1\}$ exactly n times and $\{a_2\}$ exactly $N - n$ times have the same probability, the sum is simply the total number multiplied by the probability of one of them.

d. The total number of ways of dividing N outcomes into two groups, one of size n and the other of size $N - n$ is $N!/(N-n)!n!$. For the garden pea experiment, the number of sequences with 705 whites and 224 pinks is $929!/(705!224!)$.

When an experiment has several possible distinct outcomes, a result comparable to Theorem 4.2 can be arrived at using expression (4.12).

Theorem 4.3. Multinomial Distribution

IF
1. An experiment has r possible outcomes, say $a_1, a_2, \ldots, a_r$.
2. The experiment is repeated N times, and each outcome is independent of any combination of other outcomes.
3. We let $P\{a_1\} = p_1, P\{a_2\} = p_2, \ldots, P(a_r) = p_r$, so that $p_1 + p_2 + \cdots + p_r = 1$.

THEN The probability of getting $\{a_1\}$ n_1 times, $\{a_2\}$ n_2 times, and so on, where $n_1 + n_2 + \cdots + n_r = N$ is

$$P(n_1 \text{ events } \{a_1\}, n_2 \text{ events } \{a_2\}, \ldots, n_r \text{ events } \{a_r\})$$

$$= \frac{N!}{n_1! n_2! \cdots n_r!} p_1^{n_1} p_2^{n_2} \cdots p_r^{n_r} \qquad (4.14)$$

EXERCISE 4.4. An experiment was performed with pea plants, in which two parents were crossed to get an F_1. Parent P_1 had round yellow seeds, and parent P_2 had wrinkled green seeds. The F_1 plants all had round yellow seeds. Let R = allele for round seeds, r = allele for wrinkled seeds, Y = allele for yellow color, and y = allele for green color. Assume that the genes for shape and color are independent. What can you say about dominance of R versus r; of Y versus y? In the F_2 generation, what are the probabilities of P(round, yellow), P(round, green), P(wrinkled, yellow), and P(wrinkled, green)?

EXERCISE 4.5. One estimate of the probability of a mutation at each nucleotide position in a single reproductive cycle is 10^{-8}. In an organism with 10^7 nucleotides, what is the probability that no mutation takes place? Assume that the probability for mutation at each nucleotide is the same and that they are all independent. What is the probability of at least one mutation? *Hints*: a) remember that $0! = 1$; b) if x is a number whose absolute value is very small, then $\log_e(1 + x)$ is approximately equal to x. Therefore, for very small positive x, $\log_e(1 - x)$ is approximately $-x$.

4.7. CONDITIONAL PROBABILITY

In Section 4.6, we examined combinations of independent events. Suppose, however, that we ask the question, What is the probability of a homozygous plant giving pink flowers? Now we are interested in the combination of two events, {homozygous} and {pink flowers}. These two events are not independent, since knowledge of one of them alters the probability of the other. That is, having specified in advance that the plant is to be homozygous, there are only two possibilities out of the original four that need to be considered: (C,C) and (c,c). The plant has one chance out of two, or a probability of 0.5, of having pink flowers.

If the sample space consists of equally likely outcomes, then the probability of some event A, having specified condition Q, becomes simply

$$P(A|Q) = \frac{\text{number of outcomes in the event } (A \cap Q)}{\text{number of outcomes in the event } Q} \quad (4.15)$$

That is, we have simply redefined the sample space, restricting it to outcomes compatible with condition Q. Probabilities that are conditional upon the use of such a restricted sample space are called *conditional probabilities*. The symbol $P(A|Q)$ is read, "probability of A given Q."

Generalizing equation (4.15), so that the sample space need not be an equal likelihood set, we arrive at the following definition.

DEFINITION 4.3. CONDITIONAL PROBABILITY

$$P(A|Q) = \frac{P(A \cap Q)}{P(Q)} \quad (4.16)$$

4.7 CONDITIONAL PROBABILITY

With this definition, we can find

$$P(\text{pink}|\text{homozygous}) = \frac{P(\text{pink and homozygous})}{P(\text{homozygous})}$$

$$= \frac{P(\{(c,C),(C,c),(C,C)\} \cap \{(c,c),(C,C)\})}{P\{(c,c),(C,C)\}}$$

$$= \frac{P\{(C,C)\}}{P\{(c,c),(C,C)\}}$$

$$= \frac{0.25}{0.50}$$

$$= 0.5$$

On the other hand,

$$P(\text{homozygous}|\text{pink}) = \frac{P(\text{pink and homozygous})}{P(\text{pink})}$$

$$= \frac{P\{C,C\}}{P\{(c,C),(C,c),(C,C)\}}$$

$$= \frac{0.25}{0.75}$$

Notice that if A and Q are independent, then $P(A \cap Q) = P(A)P(Q)$, so

$$P(A|Q) = \frac{P(A)P(Q)}{P(Q)}$$

$$= P(A)$$

That is, the probability of A is unaffected by any knowledge about the occurrence of Q. That is just what we should like from a concept of probabilistic independence.

Equation (4.16) is often used in a different form. It may be possible to get a direct estimate of $P(A)$ and of $P(A|Q)$ so that (4.16) would be used to get an estimate of $P(A \cap Q)$,

$$P(A \cap Q) = P(A)P(A|Q) \tag{4.17}$$

The reasoning leading to equation (4.17) can be taken a step further. Suppose we need to know the probability of some event A, but $P(A)$ is hard to estimate. Next suppose that U, the set of all possible events (the set of all subsets of the sample space), is partitioned (see Appendix B1) into the subsets $B_1, B_2, \ldots, B_n$, where

$$U = \bigcup_{i=1}^{n} B_i$$

and

$$B_i \cap B_j = \emptyset \quad \text{for } i \neq j$$

Since $\cup_{i=1}^{n} B_n = U$, it is clear that $A \cap (\cup_{i=1}^{n} B_i) = A$. Using the distributive law, we get

$$A = \bigcup_{i=1}^{n} (A \cap B_i)$$

If B_1 and B_2 are disjoint, so are $(A \cap B_1)$ and $(A \cap B_2)$; that is, since the B_i's are pairwise disjoint, so are the sets $(A \cap B_i)$. This allows us to add probabilities

$$P(A) = \sum_{i=1}^{n} P(A \cap B_i) \tag{4.18}$$

Each of the terms $P(A \cap B_i)$ may now be "factored" by equation 4.17 to give

$$P(A) = \sum_{i=1}^{n} P(B_i) P(A|B_i) \tag{4.19}$$

Equation (4.19) is known as the *formula for total probability*. It allows the calculation of the total probability of an event A from a knowledge of the probabilities of the various ways in which the event A can occur.

To take an epidemiological example, we may wish to estimate the number of people in some population who carry immunity toward a particular disease. The immune group might be broken into one group of individuals who had the disease and developed immunity and another group with natural immunity, so that

$$P(\text{immune}) = P(\text{immune} \cap \text{had disease}) + P(\text{immune} \cap \text{had disease}^c)$$

and

$$P(\text{immune}) = P(\text{had disease}) P(\text{immune}|\text{had disease})$$
$$+ P(\text{not had disease}) P(\text{immune}|\text{not had disease})$$

4.8 EXPECTED VALUES

Using this formulation, medical records or surveys may yield an estimate of P(had disease). The probability of the complement, *not had disease* is estimated as $1 - P(had\ disease)$. Each of the two subpopulations may then be sampled and tested separately to get estimates of the conditional probabilities and then $P(immune)$ calculated from them.

Equation (4.19) is heavily used in survey statistics. For example, the event A may be the event that a randomly chosen citizen votes for a particular candidate, B_i may be the event that the citizen is a member of a particular socioeconomic group. The probabilities $P(B_i)$ might be obtained from census data, and the $P(A|B_i)$ determined by surveying the voting preferences of that socioeconomic group.

In Section 4.11, equation (4.19) is applied to the garden pea example to determine genotype probabilities for generations past the F_2.

4.8. EXPECTED VALUES

For a variable which has a finite set of possible values, the expectation, or expected value, is given by equation (4.5).[2] For example, suppose we wish to calculate the expected number of dots that show up when a single fair die is cast. Each result has probability $1/6$, so the expected value is $(1+2+3+4+5+6)/6 = 3.5$. Apparently, the expected value need not be an actual realizable result. It is interpretable as the expected long-run average.

---

[2]When the set of possible outcomes (the sample space) is discrete but infinite, the same definition can be used but the sum has an infinite number of terms, and the expectation is said to exist only if the sum is well defined. If the set of possible outcomes is continuous (e.g., if the value of the variable can be any real number within some interval), then the sum is replaced by an integral and again the expectation exists only if the resulting expression is well defined—that is, if the limit that defines the integral exists. Many physical variables are thought of as being continuous. However, the accuracy with which we can measure them is limited, so that the set of possible outcomes of the measurement of any physical variable is discrete. For example, we may think of temperature as being a continuous variable, whose value may be any real number. However, if the thermometer being used measures to the nearest $0.01°$, then the set of possible outcomes of a measurement is the discrete set,...,$0.0°, 0.01°, 0.02°$,.... Of course, with a more accurate thermometer, the divisions become finer; however, it is a result of theoretical physics that there is a limit *in principle* to the accuracy with which any measurement can be made. Furthermore, the size of any physical variable must always have some bound. The conclusion is that the set of possible outcomes that result from a single measurement of a physical quantity is finite. For discussions relating to probabilities of physically measurable quantities, it is, therefore, sufficient to consider the case of finite size sample spaces.

Example. Expectation for a Binomial Variable. In Section 4.6, we developed the special case of the binomial probability distribution. It arises when an experiment has two possible outcomes and the results of succeeding experiments are stochastically independent. Note that any sample space can be partitioned (in the sense of Appendix B1) into two outcomes with respect to any possible event; that is, given any event A, the sample space can be partitioned into A and A^c. If we arbitrarily label one outcome, say the event A, by a "plus" and the other, say the event A^c, by a "minus," then we might let $P(\text{plus}) = p$ and $P(\text{minus}) = 1 - p$. It follows from equation (4.5) that the expected number of pluses from a single experiment is $0 \cdot (1-p) + 1 \cdot p = p$. For example, the expected number of pinks found upon examining a single F_2 plant is 0.75; the expected number of heads from a single flip of a fair coin is 0.5. Remember that this must be interpreted as an expected long-run average. To put it another way, the probability that is associated with an event is a numerical measure of the expectation of its occurrence.

It is frequently the case that we must deal with the expectations of sums of different variables. For example, in a certain insect population, X might be the number on the ground and Y the number on the leaves of the trees, whereas the interest might be in the total population, which is their sum, $X + Y$. In the garden pea example, X might be the number of pinks (0 or 1) in one F_2 plant, whereas Y is the number of pinks in a succeeding F_2 plant; we ask about the expected number of pinks in the two succeeding F_2 plants.

If we let $\{x_i\}$ be the possible values of X, $\{y_j\}$ the possible values of Y, and $\{z_k\}$ the possible values of $Z = (X + Y)$, we have

$$\mathcal{E}(X + Y) = \sum_{\substack{\text{values of} \\ X + Y}} z_k P(X + Y = z_k)$$

$$= \sum_{i=1}^{n_i} \sum_{j=1}^{n_j} (x_i + y_j) P(X = x_i, Y = y_j)$$

$$= \sum_{i=1}^{n_i} \sum_{j=1}^{n_j} x_i P(X = x_i, Y = y_j)$$

$$+ \sum_{i=1}^{n_i} \sum_{j=1}^{n_j} y_j P(X = x_i, Y = y_j)$$

4.8 EXPECTED VALUES

$$= \sum_{i=1}^{n_i} x_i \sum_{j=1}^{n_j} P(X=x_i, Y=y_j)$$

$$+ \sum_{j=1}^{n_j} y_j \sum_{i=1}^{n_i} P(X=x_i, Y=y_j) \quad (4.20)$$

Now we can make use of equations (4.18) and (4.19). For any *given i*, the events $(X=x_i, Y=y_1), (X=x_i, Y=y_2), (X=x_i, Y=y_3), \ldots, (X=x_i, Y=y_{n_j})$ are pairwise disjoint. Together they exhaust all the ways of getting $X=x_i$. Therefore,

$$P(X=x_i) = \sum_{j=1}^{n_j} P(X=x_i, Y=y_j)$$

Using the same argument for $Y=y_j$, we get

$$P(Y=y_j) = \sum_{i=1}^{n_i} P(X=x_i, Y=y_j)$$

Substituting these two expressions into (4.20), we get

$$\mathcal{E}(X+Y) = \sum_{i=1}^{n_i} x_i P(X=x_i) + \sum_{j=1}^{n_j} y_j P(Y=y_j)$$

But the two terms on the right-hand side are just the definitions of $\mathcal{E}(X)$ and $\mathcal{E}(Y)$. The result is

$$\mathcal{E}(X+Y) = \mathcal{E}(X) + \mathcal{E}(Y) \quad (4.21)$$

If we want the expectation of the sum of three variables, say $(X+Y+Z)$, we can let $V = X+Y$. From (4.21), we get

$$\mathcal{E}(V+Z) = \mathcal{E}(V) + \mathcal{E}(Z)$$

Applying (4.21) to $\mathcal{E}(V)$, this becomes

$$\mathcal{E}(X+Y+Z) = \mathcal{E}(X) + \mathcal{E}(Y) + \mathcal{E}(Z)$$

In this way, we can build up to any finite number of variables, which we might label $X^{(1)}, X^{(2)}, \ldots, X^{(N)}$;

$$\mathcal{E}(X^{(1)} + X^{(2)} + \cdots + X^{(N)}) = \mathcal{E}(X^{(1)}) + \mathcal{E}(X^{(2)}) + \cdots + \mathcal{E}(X^{(N)}) \quad (4.22)$$

That is, "the expectation of the sum is the sum of the expectations."
If all the expectations on the right are the same, we get

$$\mathcal{E}(X^{(1)} + X^{(2)} + \cdots + X^{(N)}) = N\mathcal{E}(X) \quad (4.23)$$

For the special case of a binomial distribution, the expected number of "pluses" in a series of N trials is, therefore, Np; the expected number of pinks in a series of $929 F_2$ plants is $(929)(0.75) = 696.75$. This has to be compared with the experimental result of 705.

4.9. VARIABILITY

Whether we are looking at a single outcome or a whole group of them, we like to have an idea not only of the expected (or average) result, but also of how close we might expect to come to that expectation. If we are throwing darts at a board, we hope that the darts are distributed around the bull's eye. If so, the bull's eye might be the expected or average position of all the darts. However, the scorekeeper (who is generally of doubtful mathematical erudition) is likely to point out that precious few of the darts got very close to the bull's eye, regardless of what their *average* position might be.

What we need is a measure of the expected amount by which we miss the expected value. For this purpose we might use $\mathcal{E}[X - \mathcal{E}(X)]$. But this is always going to be zero since

$$\mathcal{E}[X - \mathcal{E}(X)] = \mathcal{E}(X) - \mathcal{E}[\mathcal{E}(X)]$$
$$= \mathcal{E}(X) - \mathcal{E}(X)$$
$$= 0$$

We might use $\mathcal{E}[|X - \mathcal{E}(X)|]$. This would be perfectly legitimate. However, absolute values can get awfully tough to handle mathematically. In order to get something meaningful but not too hard to handle, we usually use the variance, which is defined as

$$\text{VAR}(X) = \mathcal{E}\left([X - \mathcal{E}(X)]^2\right) \quad (4.23)$$

4.9 VARIABILITY

where

$$\mathcal{E}\left([X-\mathcal{E}(X)]^2\right) = \sum_{\substack{\text{values of} \\ [X-\mathcal{E}(X)]}} [x_i - \mathcal{E}(X)]^2 P[X - \mathcal{E}(X) = x_i - \mathcal{E}(X)]$$

$$= \sum_{i=1}^{N} [x_i - \mathcal{E}(X)]^2 P(X = x_i) \qquad (4.24)$$

The square root of VAR(X) is called the *standard deviation* of X.

While this may look strange at first glance, we see in Chapter 8 that adding squares and then taking the square root is a perfectly natural way to take a total distance. It comes straight out of the Pythagorean theorem for a right triangle. The standard deviation may, therefore, be regarded as the expected distance from the mean (or expectation).

Example. ***Variance for Binomial Probability Distribution.*** If we let X be the number of pluses in a single trial, then the actual value of X can be either zero or one. The result of Section 4.8 is that the expected value for a single observation is p, the probability that $X = 1$. The variance for a single observation will be, by equation (4.24),

$$\text{VAR}(X) = [0 - \mathcal{E}(X)]^2 P(X=0) + [1 - \mathcal{E}(X)]^2 P(X=1)$$

$$= p^2(1-p) + (1-p)^2 p$$

$$= p(1-p) \qquad (4.25)$$

In Section 4.8, it was found that the expectation of a sum is the sum of the expectations. The comparable result is not true in general for variances. It can be shown that it *is* true if the events determining the variable are independent in the sense of Definition 4.2. That is, if the events determining variable X are independent of those determining variable Y, then it is true that

$$\text{VAR}(X+Y) = \text{VAR}(X) + \text{VAR}(Y)$$

In the example of the binomial distribution, subsequent outcomes *are* independent. Therefore, if we let T be the total number of pluses in N

trials, we have $\mathcal{E}(T) = Np$ and

$$\text{VAR}(T) = Np(1-p) \tag{4.26}$$

Since p is fixed, equation (4.26) says that the variance in a binomial distribution is proportional to the number of observations. The standard deviation is, therefore, proportional to $N^{1/2}$.

To show that the addition of variances is not always valid, let us look at two variables that are not independent in the sense just discussed:

X = number of pinks in a single trial

Y = number of homozygous in a single trial

$Z = X + Y$

We have

$$\text{VAR}(X) = 0.1875$$

$$\text{VAR}(Y) = (0.5) \cdot (1 - 0.5) = 0.25$$

The variable Z can take the values 0, 1, or 2. Its expectation is

$$\mathcal{E}(Z) = (0) \cdot (0.25) \cdot (0.5) + (1) \cdot (0.25) \cdot (0.5) + (1) \cdot (0.75) \cdot (0.5)$$
$$+ (2) \cdot (0.75) \cdot (0.5)$$
$$= 1.25$$
$$= \mathcal{E}(X) + \mathcal{E}(Y)$$

However, its variance is

$$\text{VAR}(Z) = (0 - 1.25)^2 \cdot (0.25) \cdot (0.5) + (1 - 1.25)^2 \cdot (0.25) \cdot (0.5)$$
$$+ (1 - 1.25)^2 \cdot (0.75)(0.5) + (2 - 1.25)^2 (0.75)(0.5)$$
$$= 0.8125$$
$$\neq \text{VAR}(X) + \text{VAR}(Y)$$

For the garden pea experiment, $\text{VAR}(X) = 0.1875$ (standard deviation = 0.433), whereas the series of 929 observations gives a standard deviation of

4.9 VARIABILITY

13.2. In the series of 929 observations, it would, therefore, not have been surprising if the observed number of pinks differed from the expected number of 696.75 by as much as 13. The observed difference was 8.25.

In evaluating the expected departure from the expected value, it is often more meaningful to express it as a percentage or fraction of the expectation. The result is called the coefficient of variation (CV),

$$CV = \frac{\text{standard deviation}}{\text{expectation}} \quad (4.27)$$

For the special case of the binomial distribution,

$$CV = \frac{(Np(1-p))^{1/2}}{Np}$$

$$= \frac{1}{N^{1/2}} \left(\frac{1-p}{p} \right)^{1/2} \quad (4.28)$$

That is, as N goes up, the standard deviation goes up, but expectation goes up faster, so the CV goes down. If N is large enough, then the expected error becomes an insignificant fraction of the mean. In such cases, we no longer need to worry about the variability. We can pretend that every one of the N elementary results is equal to its expectation and forget all about probabilities. For the garden pea experiment, the coefficient of variation (assuming the model to be correct) would be

$$CV = \frac{1}{(929)^{1/2}} \left(\frac{0.25}{0.75} \right)^{1/2}$$

$$= 0.019$$

For most purposes, a standard deviation that is 0.019 of the mean would be considered negligible. Thus we may wish simply to say that under the conditions of the Mendelian experiment, three-quarters of the F_2 plants *will* have pink flowers. If we confine ourselves to groups of, say, 10,000 plants or more, then this is a useful statement. However, if we need to look at groups of three or four plants, then the fine structure of the phenomenon becomes important, and such a deterministic assertion as "three out of four plants will have pink flowers" is not very useful.

4.10. SUMMARY OF BASIC IDEAS

Probability theory may be thought of as a model of a certain aspect of reality, a model of a certain aspect of subjective human experience. Any theory that we formulate concerning "laws of chance" must be consistent with that experience or it is rejected. To take a simple example, if a "fair" coin is tossed a thousand times, we expect to get very close to 500 heads and 500 tails—and this is true for the second "run" of 1000 tosses, for the third, and so on. If the coin is *not* fair, we may get, for example, 600 heads and 400 tails. If so, we expect on the second 1000 tosses to get not this exact result, but something very close to it. Expectations of this type are borne out by repeated experience. In the garden pea example, we have examined one series of 929 "tosses." Of course, plants are subject to more sources of variability than tosses of a coin. However, if repeated trials continue to give us close to the same result (i.e., 75% pinks and 25% whites), then the model allows us to interpret the result in terms of the occurrence of certain elementary events and the probabilities assigned to these elementary events. In a mathematical sense, these probabilities are nothing but numbers that are assigned to certain sets. The physical interpretation of these numbers results from our expectation that if the observation is independently repeated a large number of times, then the relative frequency with which a certain event occurs is closely approximated by the number we call "probability." Any theory of chance that contradicts these expectations would not be relevant to the real world, no matter how elegant the mathematics. It is remarkable that the originators of modern probability theory were able to perceive the essence of the concepts of likelihood and of chance clearly enough to capture them within the framework of a mathematical structure.

The process of forming a probability model begins with the formulation of the sample space (call it S) of all possible individual outcomes of an experiment or observation. The mathematics of probability says nothing directly about what should be in the sample space, except that the set must exhaust all possibilities and the outcomes must be mutually exclusive.

Subsets of S are called *events*, and it is to these subsets that we assign the numbers called probabilities. In order to preserve the characteristics of what we should like to call "probability," the assignment of numbers must be done according to the following simple rules or *axioms*:

a. The probability of any event (subset of S) cannot be less than zero nor greater than one.

b. Since the set S is itself an event (any set is a subset of itself—though not

4.10 SUMMARY OF BASIC IDEAS

a proper subset), it also has a probability, and

$$P(S) = 1$$

This simply means that it is certain that *something* happens.
c. If A and B are disjoint events (subsets of S), then

$$P(A \cup B) = P(A) + P(B)$$

This addition can be extended to include the union of any finite number of pairwise disjoint events.

Any assignment of numbers that meets these three criteria is a *mathematically* proper probability assignment. Only consideration of the *physical* implications of the assignment can reveal whether it is a meaningful description of a given system or set of physical events. For example, we may, after formulating the outcome set, assign to each elementary event some number that indicates an intuitive feeling about its relative likelihood. That is, we may say that event$\{c\}$ is twice as likely as $\{b\}$ and $\{b\}$ is three times as likely as $\{a\}$. Events $\{a\}$, $\{b\}$, and $\{c\}$ might then be assigned the *weights* of 1, 3, and 6, respectively. If then, each weight is divided by the total of them all (to give 0.1, 0.3, and 0.6), the relationships are preserved and the resulting numbers satisfy the three *axioms*.

A number of very useful results come directly from these axioms:

a. Since for any event A, $A \cap \varnothing = \varnothing$ and $A \cup \varnothing = A$, we have

$$P(A \cup \varnothing) = P(A) + P(\varnothing)$$
$$= P(A)$$

So,

$$P(\varnothing) = 0$$

That is, the empty set represents the "impossible" event that there is *no* outcome.
b. Since $A \cup A^c = S$ and $A \cap A^c = \varnothing$, the probability that A does *not* happen is $P(A^c) = 1 - P(A)$.
c. If $A \subset B$, then since A and $(B - A)$ are disjoint and since $B = A \cup (B - A)$, we have $P(B) = P(A) + P(B - A)$. Now, since $P(B - A)$ cannot be

negative, we conclude that

$$P(B) \geqslant P(A)$$

Now note that if $A \subset B$, then A *implies* B (if A has occurred, then B has certainly occurred). What we have shown, therefore, is that if A implies B, then B is at *least* as likely as A.

d. If E_i is the ith elementary event of a sample set with N elements, then since the E_i are pairwise disjoint and since

$$\bigcup_{i=1}^{N} E_i = S$$

we have

$$\sum_{i=1}^{N} P(E_i) = 1$$

4.11. EXAMPLE FROM MENDELIAN GENETICS; FORMULATION OF THE MODEL

Using the results of the previous sections, the model described in Section 4.2 can be translated into a set of formal probability statements. First, every F_1 is certainly heterozygous and, therefore, pink-flowered. Second, the four possible outcomes in the F_2-genotype sample space are equally likely. Next, we try to determine the probabilities in the F_3 generation, assuming that the F_2 plants are *randomly crossed* to get the F_3 generation. To begin with, we get an expression for the probability of the event $\{(c,C)$ in $F_3\}$. This is the intersection of two independent events: {receive c from female parent} and {receive C from male parent}. Therefore,

$$P\{(c,C) \text{ in } F_3\} = P\{c \text{ from female } F_2\} \cdot P\{C \text{ from male } F_2\} \quad (4.29)$$

Using the formula for total probability (4.19), $P\{c$ from female $F_2\}$ can be

4.11 MENDELIAN GENETICS; THE MODEL

expanded to give

$P\{c \text{ from female } F_2\} = P\{\text{female } F_2 \text{ is } (c,C)\} \, P\{\text{transmits } c|(c,C)\}$

$+ P\{\text{female } F_2 \text{ is } (C,c)\} \, P\{\text{transmits } c|(C,c)\}$

$+ P\{\text{female } F_2 \text{ is } (c,c)\} \, P\{\text{transmits } c|(c,c)\}$

$+ P\{\text{female } F_2 \text{ is } (C,C)\} \, P\{\text{transmits } c|(C,C)\}$

$= \dfrac{1}{4} \cdot \dfrac{1}{2} + \dfrac{1}{4} \cdot \dfrac{1}{2} + \dfrac{1}{4} \cdot 1 + \dfrac{1}{4} \cdot 0$

$= \dfrac{1}{2}$

A similar expansion shows that $P\{C \text{ from male } F_2\} = 1/2$; so equation (4.29) gives

$$P\{(c,C) \text{ in } F_3\} = \dfrac{1}{4}$$

Repeating this same reasoning for each of the other genotypes leads to the same probability, 1/4. Therefore, as long as the crossing is random (each F_2 plant has the same chance of being mated with each other F_2 plant), the probabilities in the F_3 generation are the same as in the F_2.

Reiterating for the F_4 and later generations again leads to the same set of genotype probabilities. That is, beginning with the F_2 generation, the genotype probability distribution is *stationary*; it does not change from generation to generation under the assumption of random mating.

The probability of pink flowers in a randomly chosen plant in any generation past F_1 is, therefore, 0.75. The probability of getting any *given* number of pinks out of N randomly chosen plants is given by equation (4.13). The *expected number* of pinks is $0.75N$, with standard deviation of $0.4330 N^{1/2}$. It has already been pointed out (Section 4.9) that for the series of 929 F_2 plants observed by Mendel, the results are within one standard deviation of the result predicted from the model; that is, the amount by which the result missed the expected value is well within the bounds expected by chance variability. It may be concluded that the model is consistent with the observations.

This model is a simple representation of the genetic transmission of a single qualitative phenotypic character, whose expression is not dependent upon the environment, and for which one allele is completely dominant.[3] The inputs are the genotypes of the parent varieties. The possible states and outputs of the system are the genotypes of the progeny, and the variables are the probabilities associated with these genotypes.

4.12. UNCERTAINTIES IN THE SYSTEM: STOCHASTIC VERSUS DETERMINISTIC MODELS

When the operation of a system is such that the output is uncertain, it is said to be a *stochastic* (as opposed to *deterministic*) system. When the model is structured to account for this uncertainty (or a part of it), the model is referred to as a *stochastic model*. The equations of a stochastic model do not yield a specific unique output. Rather, they yield a probability distribution over the output sample space; they associate each *possible* output with a probability.

Every real system must be considered to be subject to uncertainties of one type or another, all of which are ignored in the formulation of a deterministic model. As a result, deterministic models generally present fewer mathematical difficulties, but can only be considered to describe system behavior in some average sense. Stochastic models are required whenever it is necessary to explicitly account for the randomness of underlying events. A partial listing of situations requiring such an accounting would include the study of mutation rates, rates of evolutionary development, medical diagnosis, chromosome mapping, *in vivo* enzyme kinetics, the general theory of disease spread and epidemiology, population growth, ecological systems, radiation biology, learning behavior, and so on and on.

Connection between stochastic and deterministic descriptions is made through a consideration of expected values. It is often the case that when the number of actual events becomes very large, the coefficient of variation becomes very small, so that the observable result is indistinguishable from the expected value of the result. If this is true and if consideration is limited to large groups of events, it may be adequate to neglect any underlying uncertainties or randomness, and to use the relative mathemati-

---

[3]Many phenotypic characters, particularly those known as "continuous characters," are especially susceptible to the effects of environment, that is, weather and soil conditions, competition with other plants and so on. Such characters generally depend on the simultaneous action of more than one gene, so that the description of the elementary events is complicated by the necessity to include various types of interactions between the genes. (See Falconer, 1960; Mather and Jinks, 1971.)

4.12 UNCERTAINTIES IN THE SYSTEM

cal simplicity of a deterministic description. For example, in the garden pea model, the coefficient of variation for number of pinks in N observations is $0.57735/N^{1/2}$. The statement, "three fourths of all F_2 plants will have pink flowers," would be associated with an expected relative error (as measured by the CV) of 29% if a group of four plants is being examined; but the expected relative error for a group of 10,000 plants would be about 0.5% which would be regarded as negligible for many purposes.

There are a number of sources for uncertainties in the measureable output of a system. Some of them are fairly obvious: a) uncertainties as to the actual inputs, b) uncertainties as to the values of the parameters, and c) measurement errors. Others are perhaps more subtle: d) uncertainty as to the structure of the "true" system equations, and e) inherent randomness of underlying events. As is discussed later in this section, the uncertainties built into the stochastic model of garden pea genetics (Sections 4.2 and 4.11) may be considered to belong to either category (d) or (e).

a. Uncertainties in the Inputs. In the garden pea example, such uncertainties might be introduced if the genotypic purity of the parent lines were in doubt. In the public address system of Section 2.1, it is easy to appreciate that the overall input I might be subject to random fluctuations and uncertainties. In building a model that accounts for such uncertainties, one might specify a set of possible inputs and then assign a relative likelihood or probability to each; that is, one would specify a distribution of probabilities on the set of possible inputs. A purely deterministic description would work simply with the average or expected input. However, even when using a deterministic description, it must be realized that inputs to real world systems are nearly always uncertain to some extent. Therefore, a model that exhibits the "correct" behavior only when the input exactly equals some specific value is to be regarded with suspicion; the behavior of a deterministic model should be stable enough to allow for realistic uncertainties and fluctuations in the inputs.

b. Uncertainties in the Parameters. In the public address system of Section 2.1, the efficiencies of the microphone and of the speaker, the gain of the amplifier, the fraction being fed back, and the time delay could all be subject to unspecified environmental influences such as temperature and humidity, as well as random fluctuations sometimes called "noise." As in the case for inputs, these uncertainties may be built in to the model by specifying a distribution of probabilities on a set of possible parameter values. Also as for inputs, a deterministic description would work simply with the expected values of the parameters but should be stable enough to allow for realistic uncertainties and fluctuations.

c. Measurement Errors. This refers to uncertainties that arise in the observation of the system. At first glance, it may seem that such uncertainties are not a concern of the model builder. However, when real world systems are being described, it must be borne in mind that it is the measurement process that converts the "true" system output into an observed output. From this point of view, it is often fruitful to include the measuring process as *part* of the system to be modeled. Furthermore, the manner in which a system is observed is in large measure determined by the description, that is, the model, that has been chosen. It is desirable, *though not always possible*, that the formulation of the model be in terms of parameters, variables, and components that can be subject to easy and accurate observation. At the opposite extreme are models that contain components, variables, or parameters that are *in principle* unobservable. The usefulness of such models is obviously limited.

d. Uncertainty as to the Structure of the System Equations. It is almost never certain that the precisely correct form has been found for the system equations. Moreover, it may be that the structure of the system is itself subject to perturbing influences, which have a random character. In constructing a stochastic model that takes these uncertainties into account, the elementary events within the system may be regarded as random. In the formulation of such a model, one must specify the set of all elementary events that can occur under any given set of conditions (inputs, parameter values, and states of the system) and their relative likelihoods; that is, one must specify a distribution of probabilities on the set of elementary events that characterize the operation of the system. The resulting model equations may then accept specific inputs and parameter values, but will yield a distribution of probabilities on the set of possible outputs. The model of garden pea genetics (Sections 4.2 and 4.11) is such a model; it accepts a specific pair of parental genotypes and yields a set of probabilities on a set of progeny genotypes. When it is sufficient to describe the average behavior of the system, one may employ a deterministic model. Although such a model is not expected to account for uncertainties, its behavior should be stable enough to *accommodate* them. This type of stability is called *structural stability* and is discussed briefly in Appendix A5.

e. Uncertainty Inherent in the Operation of the System. Explicit consideration of the uncertainty of underlying elementary events presents certain

4.12 UNCERTAINTIES IN THE SYSTEM

philosophical difficulties of interpretation. The events may be regarded as: i) being inherently deterministic, but requiring more detailed information for complete specification than is available (the viewpoint adopted in the preceding paragraph) or ii) being *inherently* of a random and undetermined nature. Each of these views has had its vigorous proponents in the literature of the philosophy of science. However, even though the *interpretation* of the uncertainties may depend upon which of these views is correct, their *scientific description* does not. One may use the same options as for paragraph d without commitment to a particular philosophical view.

All sources of uncertainty must be considered when the model is to be used as a basis for prediction or of system management or when the validity of the model is being tested against data.

Testing the validity of a model essentially involves comparing: i) the discrepancy between an observed result and that predicted on the basis of the model with ii) the discrepancy that might be expected on the basis of a consideration of all sources of uncertainty. If i is much larger than ii, the conclusion would be that the observation is not consistent with the model and that the model must be held suspect. If i is of the same order of magnitude as ii or less, the conclusion would be that the observed result is consistent with the model, but *not* that the model has been proved (see Section 1.5). The topic of model testing is a central concern of the field of statistical analysis.

CHAPTER FIVE
DYNAMIC PROCESSES

Get used to thinking
that there is nothing
Mother Nature loves so well
as to change existing forms
and to make
new ones like them.
*Marcus Aurelius Antoninus
Meditations II*

In this chapter, we begin the study of dynamic systems—systems whose state (see Section 2.5) is changing with time.[1] If the state of the system is specified by the values of n state variables, it is convenient to represent it as a point in an n-dimensional space, which is called the *state space* for the system. As the system changes in time, the position of the point that represents the system also changes. The change of the system in time is, therefore, represented by the motion of a point in the n-dimensional state space. As the point moves, it traces out a path in the n-dimensional state space, which is referred to as the *trajectory* of the system. Thus the system and its progress in time can be represented by a path (trajectory) in state space—or by an equation that describes that trajectory.

If the state is one or two dimensional, then one can draw the trajectory and exhibit visually the path in state space along which the system

---

[1] Even though most of the discussion is in terms of rates of change with respect to time, the same considerations apply in discussing rates of change with respect to any other variable taken as the independent variable.

DYNAMIC PROCESSES

"moves." However, the drawing does not show how fast the system moves along this path. It is like the map of a road, along which an automobile is traveling. The map shows how the automobile may travel, but does not show how fast it moves. If we wish to show the rate of motion of a system in its state space (or of the automobile on the road), we must plot the position of the system (or of the automobile) versus time. *Time* then becomes the $(n+1)$st dimension.

Unfortunately, the ability to portray a curve in a n-dimensional space is lost when n is greater than three. The portrayal of a curve in three dimensions is inconvenient, but may be done by means of projective drawing or by the physical construction of three-dimensional curves. In four or more dimensions, we have only the mathematical description, our power of imagination, and the possibility of constructing two-dimensional projections as aids to the imagination.

An important characteristic of dynamic systems is that the state of the system at any given time t_i depends upon what the state was at a previous time t_{i-1} and upon the conditions that prevailed and influenced the direction and rate of change from t_{i-1} up until but not including the instant t_i. To take a simple example, the town in which a traveler wakes up this morning depends upon the town he was in yesterday and upon his direction and speed of travel during the previous day. Similarly, the state at time t_{i-1} depends upon the state at some earlier time t_{i-2} and so on. That is, the state at the given time t_i depends upon the state at some arbitrarily chosen starting time t_0 and the whole history of conditions that influenced the direction and rate of change up until (but not including) the given time t_i.

This characteristic of dynamic systems has two important immediate consequences. First, the state of the system at any given time t_i carries within it the *memory* of certain aspects of where it has been and the conditions that have prevailed previous to the time t_i. Second, the influence of any new input at the instant of time t_i does not affect the state *at* time t_i but affects the direction and rate of change in the time immediately following t_i. The effect of the input at t_i is not discernible until enough time has elapsed for a measurable change in the system state and in the observable system outputs; there is a *time delay* or a *time lag* in the effect of any input.

In describing the way in which a dynamic system changes—that is, in describing its motion in the state space—two kinds of distinction must be made: a) continuous versus discrete state spaces and b) stochastic versus deterministic descriptions.

a. Continuous versus Discrete State Spaces. The term "continuous state space" implies that the state variables are continuous, at least within the region of interest (Appendix B4). Many state variables of interest in biology are not continuous. For example, the number of individuals in a population (whether of plants, animals, microorganisms, or molecules) can only change by integer amounts. In such a case, as the point that represents the system moves through the state space (we often speak of the system "moving through state space"), it moves by discrete jumps rather than smoothly and continuously. As pointed out in Appendix B5, the formalism of differential equations to describe the direction and rate of motion is not available unless the variables are continuous. However, because this formalism is so convenient, discrete state spaces are often treated as though they were continuous. In subsection 5.1d we see that the validity of such an approximation requires that the sizes of the discrete jumps be small relative to the ability with which we can (or wish to) measure changes in the system. To put it another way, if the scale on which we observe the system is coarse relative to the scale on which the individual jumps occur, then the path of the system through state space may appear to be continuous—just as a curve drawn on a television screen appears to be continuous, even though it is composed of many individual dots of light.

b. Stochastic versus Deterministic Descriptions. In a deterministic description, the behavior of the system is completely determined by its state and by the specified conditions. As a result, a deterministic description of a dynamic system, and its evolution through time usually gives a description of a particular trajectory in state space. On the other hand, in a stochastic description, we have, for each state that the system can be in, a distribution of probabilities on a set of possible behaviors, that is, on the set of possible directions and rates of travel in the state space. As discussed in Chapter 4, the connection between deterministic and stochastic descriptions is made by considering the expected or average behavior. Although any real system must be considered to be subject to a variety of uncertainties, when the relative uncertainty is small compared with the need for accuracy, one may take advantage of the comparative mathematical simplicity of deterministic models.

In developing an explanatory model for a dynamic system, the purpose is to understand how the general laws that govern the behavior of the

5.1 PROCESSES WITH CONSTANT AVERAGE RATE

system arise from the laws that govern the constituent elementary events.[2] The point of view that is adopted is that these laws must always be considered to have a certain degree of random or uncertain character (see Section 4.12). This randomness has to be taken into account unless the coefficient of variation is small, which usually means that the density of events is very large.

In this chapter, we concentrate on describing individual processes that lead to changes in a system. Each process is taken to be an aggregate of elementary events, each governed by the same laws. In Chapters 6 and 7, we consider how different processes operating within the same system may interact to produce more complex types of system behavior.

5.1. PROCESSES WITH CONSTANT AVERAGE RATE

For illustration, we consider a collection of bees gathering nectar from flowers—rape flowers in particular (Free, 1970, Chapter 8). The variable to be studied is the total number of flower visits made by the foraging bees. Such a variable might be of importance, for example, as part of a study on crop pollination. Results of such studies are often reported in terms of average number of flowers visited per bee per unit time. The variable is discrete, and the events (going from one flower to another) are random, whereas reporting an average rate treats the process as being continuous and deterministic. In developing this example, we first (subsection 5.1a) indicate how one might build a model that seeks to account for the details of the behavior of each individual bee. In subsection 5.1b, the details are lumped to get a stochastic model of overall population behavior. Subsection 5.1c examines the resulting expected behavior. Subsection 5.1d makes the transition to a deterministic model. This particular process is examined from all these points of view because it is complex enough to make each point of view relevant, but simple enough to make it clear how the different descriptions relate to each other. The comparison is summarized in subsection 5.1e.

---

[2]This differs from a description of how the known behavior of a given system during a given time interval arose from the collection of precisely known elementary events. Decriptions of such detail require the observation of the system be completely comprehensive and absolutely precise and that adequate facilities be available for processing the usually formidable quantity of resulting data. Since these conditions are rarely satisfied, even an explanation of observed past behavior of a given system usually relies on drawing the relation between the laws of operation of the system and those that regulate the occurence of the constituent elementary events.

5.1a. Bee Foraging; Detailed Model

In this model, we try to account for the discreteness and randomness of the behavior of the individual bees. The rape flower has four nectaries, or nectar-containing tissues. When a bee visits a flower, it may take nectar from one or more nectaries. The nectaries are filled within 30 min of being emptied. If the population of bees in the field is high, flowers may be visited fairly often, so that a bee may find a nectary that is empty or only partly filled. When a bee finds an empty nectary, it may abandon the flower or try another nectary in the same flower. Figure 5.1 shows the process in terms of the component subprocesses. In the diagram, diamond-shaped boxes are used to represent decision processes; subsequent behavior differs depending on the specific decision taken. Each subprocess has its own set of possible outcomes and, therefore, its own set of uncertainties.

Subprocess 1. Uncertainties involve the length of time needed to select a new flower, as well as the identity of the flower selected. Both may be influenced by the presence of other bees.

Subprocess 2. As for 1, the uncertainties involve the time needed to make the choice as well as the identity of the choice. Since the nectaries are in different positions within the flower, they do not have equal likelihood of being selected.

Subprocess 3. Uncertainty involves time required, which may depend upon the position of the nectary.

Subprocess 4. This decision process has two outcomes whose likelihoods depend upon the number of other bees foraging.

Subprocess 5. Uncertainty involves time required, which depends upon amount of nectar, which in turn depends upon the number of other bees foraging.

Subprocess 6. Conceivably, the relative likelihoods of the two possible decisions depends upon conditions that the bee has found at the flower being visited and upon the probability of approach of another bee.

Any of these subprocesses could be influenced by time of day, air tempera-

5.1 PROCESSES WITH CONSTANT AVERAGE RATE

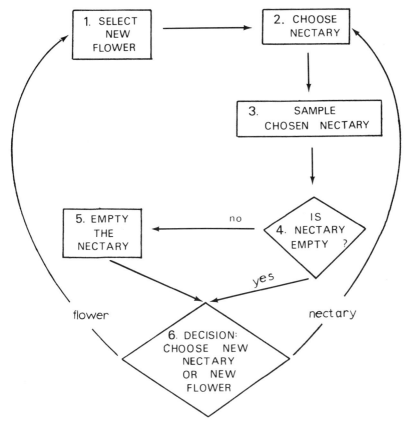

Figure 5.1. Decomposing the process of an individual bee foraging on rape flowers.

ture, wind currents, and so on, or by coordination between the behavior of individual bees.

Although it might be possible to describe such subprocesses by a set of probabilistic equations, actually doing so requires a detailed study of each. Once accomplished, the mathematical difficulties in using the resulting equations to obtain a description of the overall process would be formidable even for this comparatively simple case. Two other options are available. One of these involves computer simulation, which is discussed in Chapter 9. The other involves reducing the amount of detail being considered and is the approach of the following subsections.

5.1b. Bee Foraging; Stochastic Formulation

In this subsection, we focus directly on the variable of interest—the number of flower visits. The various subprocesses of Figure 5.1 are lumped into the one event that leads to a change in this variable, a bee landing on a new flower.

The strategy is to select a small time interval Δt and to determine the probability that some bee will land on a new flower during that interval. If we let N_L be "total number of landings," then the probability of exactly one new landing in the interval from t to $t+\Delta t$ can be expressed formally as $P(N_L = n+1$ at $t+\Delta t | N_L = n$ at $t)$. Two assumptions are initially made concerning behavior in the selected time interval.

Assumption a. The expectation of a new landing in the time interval does not depend upon what happened in any previous time interval.

Assumption b. As the length of the time interval is decreased, the expectation of a new landing during the time interval decreases; the expectation goes smoothly and continuously to zero as the length of interval approaches zero.

The implications of these assumptions are: first, if the interval is chosen small enough, the probability of more than one landing in the interval becomes negligably small; second, providing Δt is small enough, the probability of a single landing can be expressed as being approximately proportional to Δt. Using the "little oh" notation of Appendix C and denoting the proportionality constant by $k(t)$ to indicate that it might depend upon when the interval starts (e.g., 9:00 versus 12:00)[3], the following formal statements result:

$$P(N_L = n+1 \text{ at } t+\Delta t | N_L = n \text{ at } t) = k(t) \cdot \Delta t + o(\Delta t) \quad (5.1a)$$

$$P(N_L > n+1 \text{ at } t+\Delta t | N_L = n \text{ at } t) = o(\Delta t) \quad (5.1b)$$

$$P(N_L = n \text{ at } t+\Delta t | N_L = n \text{ at } t) = 1 - k(t) \cdot \Delta t + o(\Delta t) \quad (5.1c)$$

The last equation follows from the requirement that the probabilities of all possibilities add to one.

---

[3]Indeed, bees have been shown to be sensitive to polarized light, which varies during the day and vanishes in summer at midday (see Wellington, 1976).

5.1 PROCESSES WITH CONSTANT AVERAGE RATE

An important implication of assumptions a and b and of equations (5.1) is that the behavior of the individual bees is assumed not be synchronized. If, for example, the population consists of ten bees whose behavior is perfectly synchronized, then during each time interval, we would have either ten new landings or none. If these landings occurred in the immediately preceding interval, the probability of a repeat in the current interval would be affected, contrary to assumption a. Indeed, in such a case the interval length could be made short enough to make landing in two successive intervals physically impossible. Synchronization of populations is discussed further in Section 5.8.

Equations (5.1) comprise a statement about how fast the system moves in the state space. The state space has only the single dimension N_L. Motion is always in the forward direction, toward increasing values of N_L. From equations (5.1), we can obtain a description of how the probabilities $P(N_L = n)$ change for any value of n as well as a description of the expected rate of travel of the system through the state space, that is, of the *expected* rate of increase of N_L. The description of how the probabilities change is in terms of the derivatives $dP(N_L = n)/dt$, where n is any integer that can be specified. The strategy that is used to develop equations for these derivatives is typical of the strategy used in the study of stochastic models. We note first that the event that $N_L = n$ at a particular time t can arise if: i) $N_L = n - 1$ at time $t - \Delta t$ and a single landing has occurred; or ii) $N_L = n$ at time $t - \Delta t$ and *no* new landing has occurred. These two possibilities are mutually exclusive and exhaustive, so the formula for total probability (4.19) may be used to obtain

$$P(N_L = n \text{ at } t) = P(N_L = n - 1 \text{ at } t - \Delta t) P(N_L = n \text{ at } t | N_L = n - 1 \text{ at } t - \Delta t)$$
$$+ P(N_L = n \text{ at } t - \Delta t) P(N_L = n \text{ at } t | N_L = n \text{ at } t - \Delta t)$$

Using equations (5.1), this becomes

$$P(N_L = n \text{ at } t) = P(N_L = n - 1 \text{ at } t - \Delta t) k(t) \cdot \Delta t$$
$$+ P(N_L = n \text{ at } t - \Delta t)(1 - k(t) \cdot \Delta t)$$
$$+ o(\Delta t)$$

If the second term on the right is expanded and the equation is rearranged slightly, we get

$$\frac{P(N_L=n \text{ at } t)-P(N_L=n \text{ at } t-\Delta t)}{\Delta t} = k(t)P(N_L=n-1 \text{ at } t-\Delta t)$$

$$-k(t)P(N_L=n \text{ at } t-\Delta t)$$

$$+\frac{o(\Delta t)}{\Delta t}$$

If we take the limit as $\Delta t \to 0$, the left-hand side becomes $dP(N_L=n)/dt$; the Δt's vanish from the first two terms on the right-hand side; and the last term goes to zero as a consequence of the definition of $o(\Delta t)$. Therefore, for any nonnegative inter n,

$$\frac{dP(N_L=n)}{dt} = k(t)\bigl[P(N_L=n-1)-P(N_L=n)\bigr] \tag{5.2}$$

Equation (5.2) is really an infinite set of equations, one for every possible value of n. To go from (5.2), which tells how the probabilities change with time, to an expression that tells what the probabilities *are* at any given time requires that we specify how the parameter k varies. Under the assumption that k is constant with respect to time and that we start at $t=0$ with $N_L=0$, the following expression results (see Karlin, 1975 for details of the solution),

$$P(N_L=n \text{ at time } t) = \frac{(kt)^n}{n!}e^{-kt} \quad n=0,1,2,3,\ldots \tag{5.3}$$

Equation (5.3) specifies a whole probability distribution for any time t and shows how that probability distribution changes as a function of time. It is called the *Poisson* distribution and is depicted in Figure 5.2. Examination of Figure 5.2 together with equations (5.2) indicates what happens to the distribution as time goes on. On the left-hand side of the maximum, we have $P(N_L=n-1) < P(N_L=n)$, so the derivative is negative; the reverse is true on the right-hand side of the maximum. Therefore, as time goes on the probability hill builds up on the right and shrinks on the left, so that the maximum moves to the right. This, of course, is just what one should expect.

5.1 PROCESSES WITH CONSTANT AVERAGE RATE

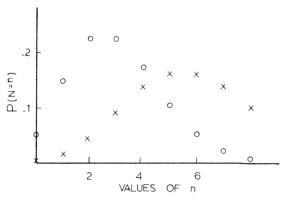

Figure 5.2. Poisson distribution given by equation (5.3) for $k=3$ and $t=1$ (circles) or $k=3$ and $t=2$ (crosses).

5.1c. Expected Behavior and Variance

The expected behavior of N_L can be obtained for constant k by mathematical manipulation of equation (5.3). However, it is both easier and more instructive to use the basic model equations (5.1) directly. The approach is to determine the amount by which the expectation changes in a small interval Δt, to use this to obtain a differential equation that describes the rate of change, and finally to use the differential equation to obtain a general expression for $\mathcal{E}(N_L)$ as a function of time.

From equation (5.1), we can calculate

$$\mathcal{E}(\text{increase in } N_L \text{ during } \Delta t) = k(t)\cdot \Delta t + o(\Delta t)$$

Noting that the increase in N_L is the difference between its value at the end and at the beginning of the interval and also that the expectation of this difference is the difference of the expectations, we obtain

$$\mathcal{E}\left[N_L(t+\Delta t)\right] - \mathcal{E}\left[N_L(t)\right] = k(t)\cdot \Delta t + o(\Delta t)$$

Dividing both sides by Δt and taking limits as $\Delta t \to 0$,

$$\frac{d\mathcal{E}(N_L)}{dt} = k(t) \tag{5.4}$$

If we again make the assumption that the parameter k does not depend upon time, this leads to

$$\mathcal{E}[N_L(t)] = kt \tag{5.5}$$

An expression for the variance can be obtained using the same strategy. Although it involves slightly more algebra than for the expectation, it follows from equations (5.1) and the definition of variance (4.24) that

$$\text{VAR(increase in } N_L \text{ during } \Delta t) = k(t) \cdot \Delta t + o(\Delta t) \tag{5.6}$$

This is the same expression as for the expectation. It leads to the same differential equation, with the same solution for constant k. (Note that because the solution involves integrating the derivative, and because an integral is the limit of a sum, the solution depends explicitly on assumption a, that the behavior during any interval of time is independent of behavior during other time intervals.) The result is that the variance is equal to the expectation and continues to increase with time. However, the *relative uncertainty* depends on the relation between the standard deviation and expectation. As measured by the coefficient of variation, the relative uncertainty is $1/(kt)^{1/2}$. It decreases with time.

The uncertainty just discussed is completely apart from any errors in the determination of the value of the parameter k; it is due purely to the randomness of the individual elementary events.

5.1d. Deterministic Model

A deterministic formulation for this process would be based on the average (i.e., expected) rate. Using the same symbols as for the stochastic model, a similar differential equation is arrived at

$$\frac{dN_L}{dt} = k(t) \tag{5.7}$$

For constant k, the value of N_L at any time t is predicted to be kt, in exact agreement with the expectation of the stochastic model. In Section 5.7, it is seen that the deterministic model does not always agree with the expectation of the corresponding stochastic model.

Equation (5.7) may be thought of as arising from (5.4) by dropping the expectation sign. Equation (5.4) is a sensible statement about the rate of change of an expectation; its correctness depends upon the correctness of the assumptions made in its derivation. However, equation (5.7) is strictly

5.1 PROCESSES WITH CONSTANT AVERAGE RATE

speaking meaningless; it is a statement about the derivative of a variable that can only have integer values and, therefore, has no derivative (see Appendix B5). Nevertheless, (5.7) provides a simple description of the process and might, under appropriate circumstances, be an adequate approximation. Some general guidelines as to what the appropriate circumstances are can be arrived at by looking at the expression $N_L(t+\Delta t) - N_L(t)$.

This quantity is the number of elementary events that have occurred in the time interval Δt. In fortunate circumstances, as for the present example, this number has a coefficient of variation that gets small as the number of events gets large. In such cases, if the number of events in Δt is large enough, then the difference between the actual number and its expectation may not be of any consequence; the value of $[N_L(t+\Delta t) - N_L(t)]/\Delta t$ is adequately approximated by its expectation. Thus assuming that Δt is long enough to accomodate a sufficiently large number of events, we may arrive at the deterministic expression

$$\frac{N_L(t+\Delta t) - N(t)}{\Delta t} = k(t)$$

Now a problem arises if we wish to take the limit as Δt goes to zero so as to obtain a differential equation. As Δt gets smaller, so does the number of events within the interval Δt. Eventually, we must get to values for Δt during which a "large" number of events does *not* occur, and the expression whose limit is being taken is no longer an adequate description of the process. In arriving at the deterministic expression (5.7), therefore, some special restrictions must be imposed.

When the limit of any mathematical expression is taken, it is asserted that there is some value L, to which the expression keeps getting closer. Thus in taking the limit to get equation (5.4), we assert the existence of a value that the ratio $\mathcal{E}[N_L(t+\Delta t) - N_L(t)]/\Delta t$ gets closer and closer to as Δt goes to zero. If this ratio were to be constant, as it would be if $k(t)$ were constant, then the ratio would already have its limiting value long before Δt gets near zero, that is, long before Δt gets so small that the number of events in the interval is no longer "large."

If $k(t)$ is not constant, then the value of the ratio may continue to change as Δt gets smaller. However, if it is changing slowly enough, the ratio becomes *approximately* constant, while Δt is still large enough to accomodate a "large" number of events. To put it another way, if $k(t)$ is not changing too rapidly with time, it will be possible to divide the time axis into intervals that are *small enough* so that the expected rate remains essentially constant within each time interval, but *large enough* to accommodate a sufficient number of events to make the coefficient of variation small. We need not actually find such intervals, but in order for the deterministic equation (5.7) to be applicable, such intervals must *exist*. They can only exist if the time scale on which the rate of the process changes is large compared to the time scale on which the elementary events take place, that is, *if the expected rate changes slowly compared with the expected time between elementary events*. Furthermore, even when the description of equation (5.7) is applicable, it must not be asked to describe changes that take place on a time scale comparable to the expected interval between elementary events. For example, if the expected time between landings is 0.1 sec, a statement as to the change in N_L over a period of less than 0.5 sec doesn't mean very much except as a probabilistic statement of expectation.

5.1e. General Remarks

In subsection 5.1a, Figure 5.1 sets out the assumptions as to the details of the process. In going to the model of subsection 5.1b, none of this detail is relevant except that it provides guidelines for a new set of assumptions—in particular that the process proceeds by discrete steps, that the probability of a step is continuous with respect to time (assumption *b*), and that the probabilities associated with different time intervals are independent of each other (assumption a). Moreover, in assuming that the parameter *k* is constant, we have assumed that the probabilities associated with different time intervals of the same length are identical to each other. All of these assumptions are retained in developing the expression for expectation in subsection 5.1c and in the deterministic model of subsection 5.1d. In subsection 5.1d, we introduce the additional restrictions that the rate of the process not be changing too rapidly and that the model not be used to describe changes over time intervals that are too short. These restrictions are further discussed in Section 5.2. In obtaining the solutions represented by equations (5.3) and (5.5), it is assumed that the rate remains constant over time.

None of the equations say anything about how the rate depends on the size of the foraging bee population. This requires an additional statement (i.e., assumption) about the process. Suppose we let $k_1(t)$ be the value of the rate parameter for a foraging "population" consisting of a single bee. Then substituting k_1 for k, all the statements of this section become statements about the behavior of a single bee. What we need next is a

5.1 PROCESSES WITH CONSTANT AVERAGE RATE 113

statement about how the bees interact. We use the simplest possible assumption; they do not interact.

Assumption c. The bees behave identically and independently of each other in a stochastic sense.

As a result, each individual bee may be described as if it were a population with the same parameter $k_1(t)$, and the expectations and variances the collection of N_B bees in the sum of the expectations and variances for the individual bees. It follows that $k(t) = k_1(t) \cdot N_B$; that is, the behavior is proportional to the number of bees in the foraging population. In the introduction to this section, it is stated that the results of bee foraging studies are often reported in terms of an average rate per bee. If such a figure is to be used as an indication of the behavior of a foraging collection of bees, assumption c is required. The discussion of the individual subprocesses in subsection 5.1a indicates, however, that the validity of the assumption may be in doubt.

EXERCISE 5.1. On what kind of time scale would it be reasonable to use a derivative formulation to describe the following?
a. Molecular changes.
b. Microorganism growth.
c. Change of size in population of rabbits.
d. Change in size of a population of elk.
e. Density of stars in the milky way.
f. Change of number of dollars in the bank.

EXERCISE 5.2. For the system you were asked to begin modeling, which relationships involve dynamic processes? (If your model system does not involve any dynamic processes, you should at this point begin work on a new model that does.) For each of these dynamic processes, is the process: (a) continuous or does it involve discrete events? (b) stochastic or deterministic? Which types of description would be appropriate: that of subsection 5.1a (detailed decomposition into stochastic subprocesses), subsection 5.1b (stochastic formulation as a single process), subsection 5.1c (stochastic formulation considering only the expectation and variance), or subsection 5.1d (continuous, deterministic formulation)? Can you make assumptions comparable to that of a and b of subsection 5.1b or to c of subsection 5.1e? If a continuous deterministic description is appropriate, on what time scale would it be appropriate?

5.2. LINEAR RATE LAWS

We continue to look at systems simple enough to be described by a single-state variable. The term *linear rate law* (equivalently, linear differential equation) implies that the rate of change of the variable is proportional to its own value.

The example that serves as the basis for initial discussion is the growth of a bacterial culture at a fixed level of nutrient; the variable of interest is the size of the bacterial population. The concepts that are developed in the discussion of this example hold for populations of other organisms, as well as for populations of a more general nature (see Section 5.5).

The approach is similar to that of Section 5.1 in that we start with a description in terms of discrete random events and lead to a description in terms of a continuous deterministic process. Discussion of the stochastic model is limited to a consideration of the behavior of the expected value of the population size.

The system to be discussed might be taken to be the isolated growth chamber of the chemostat of Appendix A2. This would be a single physical compartment, but we can represent it as comprising three conceptual compartments as in Figure 5.3. According to the diagram, build-up of waste is assumed to have no effect on intake or use of nutrients.

The size of the population is expressed in terms of number of bacterial cells. This number changes if any individual cell divides or if any individual cell dies. The assumptions we make about each of these kinds of events are closely analogous to the assumptions made in Section 5.1, though the order is different.

Assumption a. If an interval of time Δt and a single bacterial cell are selected at random, there is a certain expectation or probability that the cell will divide during the interval. The same probability is assumed to apply to all cells. If we pick several cells at the same time, the probability that a given one of them divides does not depend upon what the others do. That is, with respect to probability of division, the cells are assumed to be identical and independent.

Assumption b. The expectation that a cell will divide in an interval Δt depends only on the length of the interval and not at all on when the interval starts.

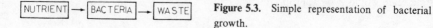

Figure 5.3. Simple representation of bacterial growth.

5.2 LINEAR RATE LAWS

Assumption c. As the length of the time interval is decreased, the expectation of division decreases, going smoothly and continuously to zero as the length the interval goes to zero.

These assumptions could not be considered to apply to the study of single bacterial cells about which we accumulate detailed histories. The uncertainties expressed by the assumptions arise because we are observing the collection as a whole and do not have—or do not wish to make use of—detailed information about individuals. Similarly, the group of assumptions would not apply to a synchronous population. In practical terms, assumptions a through c imply a homogeneous, randomly dividing population.

Using arguments analogous to those of Section 5.1 and letting λ be the "probability per unit time" parameter for division of a single cell, we obtain

$$\mathcal{E}(\text{"births" from a single bacterium in time } \Delta t) = \lambda \cdot \Delta t + o(\Delta t)$$

and for the population of N cells

$$\mathcal{E}(\text{total "births" in } \Delta t) = \lambda N \cdot \Delta t + o(\Delta t)$$

Recall that the notation $o(\Delta t)$ indicates that the expectation is approximately equal to $\lambda N \cdot \Delta t$ and that the approximation becomes better as Δt becomes smaller, finally becoming precise as Δt approaches zero.

If we now make comparable assumptions about the death of the cells and let μ be the parameter for cell death, it follows that

$$\mathcal{E}(\text{total deaths in } \Delta t) = \mu N \cdot \Delta t + o(\Delta t)$$

The number by which the population size changes in time Δt is simply births minus deaths;

$$N(t + \Delta t) - N(t) = \text{number of births} - \text{number of deaths}$$

Therefore,

$$\mathcal{E}\left[N(t + \Delta t) - N(t)\right] = N(t) \cdot (\lambda - \mu) \cdot \Delta t + o(\Delta t)$$

If the value of N is itself uncertain, this can be written,

$$\mathcal{E}\left[N(t + \Delta t) - N(t)\right] = \mathcal{E}\left[N(t)\right] \cdot (\lambda - \mu) \cdot \Delta t + o(\Delta t)$$

Dividing by Δt, taking the limit as Δt approaches zero, and letting $\kappa = (\lambda - \mu)$ yields the following differential equation for the expected value of N:

$$\frac{d\mathcal{E}[N(t)]}{dt} = \kappa \mathcal{E}[N(t)] \tag{5.8}$$

This is a linear differential equation that says that the rate of change of the expectation for N is proportional to its own value. It is often replaced by an equation of the type,

$$\frac{dN(t)}{dt} = \kappa N \tag{5.9}$$

As was the case in Section 5.1, use of the deterministic statement of equation (5.9) implies acceptance of all the assumptions that went into the stochastic statement of equation (5.8), plus the additional restriction that the rate of change of the derivative to be slow relative to the rate at which the underlying elementary events occur. When the derivative is proportional to the size of the population, this restriction usually implies that the size of the population is large relative to the rate parameter κ. This would allow an interval of time to be found in which the number of events is: i) large enough to make the coefficient of variation small and ii) small enough relative to the number of individuals in the population so that the derivative is essentially constant during the interval.

The meaning of this restriction is often difficult to grasp, but it is important to do so. As pointed out in Appendix B5, the derivative of a function of time is itself a function of time; the derivatives in equations (5.8) and (5.9) specify a value for every value of t. Since $\mathcal{E}[N]$ is a continuous variable, it makes sense to talk of its rate of change (i.e., of the slope of its time curve) at any point in time. The value of N itself, however, changes only by discrete jumps. At any instant of time it is either: i) not changing, so its time curve is flat and the rate of change is zero, or ii) jumping from one value to another, so its time curve is vertical, and its rate of change is infinite. Therefore, we cannot speak of its rate of change at an instant of time; we can only speak of an average rate of change within an interval of time. However, in order for us to do so, the average rate has to remain constant long enough to accumulate a number of events sufficient to make an average value meaningful. It follows that even when the restriction is satisfied, the resulting deterministic rate equation must *not* be asked to describe changes in the population over intervals of time that are small enough to be comparable to the average time between underlying events.

Even though all of the discussion in this and the previous section has been in terms of rates of change with respect to time, the same considerations apply in discussing rates of change with respect to any other kind of variable. We might be discussing the way in which the density of any entity (chemical compound, biological species, etc.) varies with geographical location. Instead of discussing number per unit of time, we would need to discuss number per unit volume or area and the way in which this ratio changes as we move from one region to another.

5.3. BEHAVIOR OF SIMPLE LINEAR SYSTEMS

Linear differential equations of the type of (5.9) are often used to describe changes in real world systems. They are relatively simple to handle mathematically and reflect relatively simple assumptions as to the population that is changing: that individuals are stochastically identical and independent. For these reasons, such equations are often used as a first approximation even when it is realized that the assumptions are not satisfied and that the description will later have to be modified.

So that the discussion in this section does not seem bound to any particular example, we let y denote the variable that is changing. We assume that y is continuous or that the appropriate time scales can be found so that it may be treated as if it were continuous. From the discussion of the previous two sections, we may reason that the differential equation

$$\frac{dy}{dt} = \kappa y \qquad (5.10)$$

applies when:

a. The creation or destruction of y (or the physical entity that y represents) during any time interval Δt depends only on how much y there is at the beginning of the time interval and how long the interval is, but not on when the interval is chosen (homogeneity with respect to time);
b. The physical entity that y represents may be considered to consist of increments which contribute to the creation or destruction of y equally and independently (at least in an average or probabilistic sense).

In such a case, a plot of the rate of change of y (the derivative of y) against y itself is a straight line whose slope depends on κ. If y represents an actual physical entity, it is usually restricted to being either positive or zero. The case of $y = 0$ is not especially interesting; the rate of change would also be zero, and y would remain equal to zero forever. If $y > 0$, then the derivative is positive or negative, depending on the sign of κ. In the example of bacterial growth, this depends on the relative magnitudes of the birth–death parameters.

EXERCISE 5.3. In the system you are modeling, which dynamic processes can be described by linear rate laws?

Before looking at the explicit solution to the differential equation, let us attempt to reason our way to an understanding of the progress of the population. Suppose κ is positive and we start off with an initial value of y, call it y_0, which is positive. In this case, the rate of change is positive, so that after a short period of time, y has increased. But since the rate of change is proportional to the amount, the rate of change also has increased. A curve that shows y as a function of time, would thus have to be a curve with an ever increasing slope, such as in Figure 5.4a. As y gets large, the slope gets larger. It approaches infinite slope as y continues to increase; that is, the curve approaches a vertical line. Of course, if y represents an actual physical entity, its amount can never truly become infinite. Therefore, as y gets big enough, a curve such as in Figure 5.4a cannot continue to describe the system. Working backwards through the reasoning process, we find that the rate law, $dy/dt = \kappa y$ cannot continue to apply and that the underlying assumptions about the system cannot continue to hold. In the case of microorganism growth, we indeed find that eventually the available nutrient is depleted and that toxins are accumulated, so that the value of the parameter κ decreases. Eventually it may become negative.

In the case of negative κ and a starting value y_0, which is positive, the rate of change is negative, so that y is decreasing. As y gets less, its rate of decrease likewise becomes less. As y gets closer and closer to zero, the slope comes closer and closer to zero. We wind up with a curve of continuously *decreasing* slope; it approaches zero slope as y approaches zero (Figure 5.4b). Here again, we cannot expect the equation to continue to apply forever, unless y is truly a continuous variable. Eventually, y becomes so small, that its discrete nature has to be considered. This may be appreciated by considering the bacterial growth example, when $N = 1$. The death of a single organism abruptly terminates the process.

5.3 BEHAVIOR OF SIMPLE LINEAR SYSTEMS

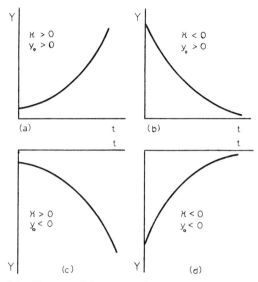

Figure 5.4. The type of time curves that can result from equation (5.10).

If the initial value y_0 is negative (this would usually mean that y represents some abstract conceptual entity), then the qualitative behavior for $\kappa > 0$ and $\kappa < 0$ become "mirror images" of that for y_0 positive. In this case, if $\kappa > 0$, then y becomes more negative as time goes on, and its rate of doing so gets greater (Figure 5.4c). If $\kappa < 0$, then y gets less negative as time goes on, and its rate of doing so decreases as $|\kappa y|$ decreases (Figure 5.4d).

In elementary calculus, it is found that when a function has a derivative given by an equation such as (5.10), the function has the form

$$y(t) = y_0 e^{\kappa(t-t_0)} \qquad (5.11)$$

where t_0 is the starting or initial time and y_0 is the starting value of y, called its *initial value*. Equation (5.11) is not the equation for any specific curve until we specify values for the parameter κ, the initial value y_0, and the initial time t_0. Once these values are specified, we obtain a curve that has the qualitative shape of one of the curves of Figure 5.4 (assuming $\kappa \neq 0$ and $y_0 \neq 0$). It is frequently convenient to arbitrarily let t_0 be zero, so that (5.11) becomes

$$y(t) = y_0 e^{\kappa t} \qquad (5.12)$$

Equations (5.11) and (5.12) are descriptions of *exponential change* that arise from a *linear rate of change*. These descriptions are based on the assumptions of the stochastic independence and identicalness of the underlying increments of y and of homogeneity with respect to the time structure in which the system operates.

If we take logarithms (base e) of equation (5.12), we get

$$\ln y(t) = \ln y_0 + \kappa t \qquad (5.13)$$

Equation (5.13) is useful for testing the description of the system as well as for determining the values of κ and y_0. The value of y is measured at certain values of t, and a plot of $\ln y$ versus t is constructed. If the description is correct, the plot is a straight line, the slope is the parameter κ, and the intercept is $\ln y_0$. If the plot is not linear and the deviation from linearity cannot be assigned to experimental error, then something is wrong with the description and with the assumptions upon which the description is based. A conclusion that "something is wrong" must also be accepted if the values obtained for κ and y_0 are unreasonable in the light of other knowledge that we have about the system.

5.4. CHARACTERIZING THE RATE OF EXPONENTIAL CHANGE

The rate of a linear process (associated with exponential change) may be completely described by quoting the value of the "rate constant" κ of equation (5.10). This parameter has dimension $(\text{time})^{-1}$ (the reader should verify this if it is not immediately obvious). Often, however, physical and biological intuition is better served by using a parameter that has dimension (time). Such a parameter usually gives the time required for a certain proportion of change to take place.

NOTATION. It is often considered preferable to assume that the rate parameter is positive and to explicitly exhibit the negative sign if necessary. In this discussion, it is convenient to let k denote positive parameters and to use the letter κ in general discussion when the sign is deliberately left unspecified. With this notation, equation (5.10) splits into the two cases (assuming $t_0 = 0$ and $y_0 > 0$),

$$\text{decreasing population:} \quad \frac{dy}{dt} = -ky; \quad y(t) = y_0 e^{-kt} \qquad (5.14)$$

$$\text{increasing population:} \quad \frac{dy}{dt} = ky; \quad y(t) = y_0 e^{kt} \qquad (5.15)$$

5.4 CHARACTERIZING THE RATE OF EXPONENTIAL CHANGE

5.4a. Half-Life for Exponential Decrease

The half-life is one of the parameters most frequently used to describe exponential decrease. It is the time required for the amount to be reduced to half of the initial amount, and is easy to find in terms of the rate constant.

Starting at time $t_0 = 0$, there is a time $t_{1/2}$ at which y is half its starting value; that is,

$$y(t_{1/2}) = \frac{1}{2} y_0$$

Substituting $t_{1/2}$ into equation (5.14),

$$y(t_{1/2}) = \frac{1}{2} y_0 = y_0 e^{-kt_{1/2}}$$

Using logs to base e, that is easily solved for $t_{1/2}$,

$$t_{1/2} = \frac{1}{k} \ln 2 \qquad (5.16)$$

Note that $t_{1/2}$ depends only on the rate parameter k, which is taken to be constant. The time needed to reduce the initial amount to half does not depend on the initial amount. Therefore, the time to remove half of what is left would again be this same $t_{1/2}$. Starting from any point on the curve (Figure 5.4b), we get the same half-life. To say that half of what remains is always removed in the same period is to say that the absolute rate of decrease is decreasing, but the *relative* rate $(1/y) \cdot (dy/dt)$ is constant; this is nothing but a restatement of equation (5.10). Therefore, constancy of half-life is a unique characteristic of linear rate processes.

5.4b. General Fractional Life; Relaxation Time

While $t_{1/2}$ often is an intuitively appealing way to describe exponential decrease, we could use the time required for $y(t)$ to be reduced to any fraction of y_0 that we choose. Suppose we designate by t_f the time required so that

$$\frac{y(t_f)}{y_0} = f; \qquad 0 < f < 1 \qquad (5.17)$$

Then, proceeding in exactly the same way as for equation (5.16), we get

$$t_f = \frac{1}{k} \ln\left(\frac{1}{f}\right) \qquad (5.18)$$

(Unless this result is completely obvious, the reader should derive it step by step.)

An often used value for f is $1/e$, where e is the base of the natural logarithms. This is convenient because $\ln e = 1$, so that

$$t_{1/e} = \frac{1}{k} \qquad (5.19)$$

The value $t_{1/e}$ is often called the *relaxation time*. The term comes from the idea that once y is reduced to zero, the system is at "rest"; it no longer is changing. Then $t_{1/e}$ is the time required for the system to come "close" to its resting point in state space.

5.4c. Doubling Time for Exponential Increase

For an exponentially increasing process, we let t_q be the time required for y to be q times its initial value,

$$\frac{y(t_q)}{y_0} = q; \qquad q > 1$$

Using equation (5.15), we get

$$y(t_q) = q y_0 = y_0 e^{k t_q}$$

so that

$$t_q = \frac{1}{k} \ln q \qquad (5.20)$$

This equation should be compared with (5.18).

If $q = 2$, we get the *doubling time*,

$$t_2 = \frac{1}{k} \ln 2$$

[compare with equation (5.16)]. Doubling time is often used to characterize

5.4 CHARACTERIZING THE RATE OF EXPONENTIAL CHANGE

the rate of increase of uncontrolled reproducing populations such as those of bacteria, people, and scientific publications. The value of $\ln 2$ is 0.69315, or approximately 0.7. Therefore, if k is expressed as percent per unit time, we have $t_2 \cong 70/k$. This is known as the "rule of seventy" and is a frequently encountered approximation for relating doubling time to the rate constant.

Just as the half-life and relaxation time for exponential decrease, so the doubling time for exponential increase is constant and does not depend upon the initial amount. Frequently made statements to the effect that the population of the world is exponentially increasing with a constantly decreasing doubling time are, therefore, either self-contradictory or are based on a loose and imprecise use of the term "exponential increase."

5.5. EXAMPLES OF SIMPLE LINEAR RATE PROCESSES

In Section 5.2, we discussed the use of a linear rate law to describe the growth of a bacterial population. Exactly the same discussion can be applied to a population of any biological organism, providing the population is homogeneous and both reproduction and death are random with respect to time. In the examples of this section, we examine the application of equation (5.10) to other types of processes.

5.5a. Unimolecular Chemical Transformations

In a unimolecular chemical reaction, an individual molecule of one compound is converted into something else. During any interval of time, a molecule has a certain probability of reacting and all molecules of the same compound are assumed to have stochastically independent and identical behavior. The result is a "pure death" process, in which the expected number of remaining molecules follows an equation of the type of (5.8), with negative rate parameter. Furthermore, the scale on which individual chemical events occur is such that unless there is a need to account for changes that occur within ultrashort time intervals (say, 10^{-6} sec or less) and in volumes so small that they contain only a few molecules, the continuous deterministic formulation of (5.9) is usually adequate.[4] A

---

[4]When considering *in vivo* biochemical transformations, the structure and organization of the cell is such that reactions may take place in isolated volume elements that contain only a few molecules and for which random fluctuations may be of significance (see, for example, Smeach and Gold, 1975).

special, but very important, type of unimolecular transformation that may be described in this way is the decay of radioactive elements. Multimolecular transformations are used in Section 5.7 as the prototype example for discussion of processes that require interactions between individuals in a population.

5.5b. Diffusion (Random Migration) Across a Barrier

Diffusion processes are extremely important in the operation of biological systems. They are involved, for example, with the transport of metabolites through membranes and through the cytoplasmic fluid, and with the rate at which substances become distributed through the blood system. Important questions in drug research concern the rate at which a given drug gets to the site of action, and how long it stays there before being "washed out." Important questions in environmental control are concerned with the rate at which pollutants diffuse through air and water; these diffusional rates must be considered along with the specific water and air currents that are found. The mathematical description of diffusion is based on the assumption that the underlying units are moving in a random fashion. Therefore, although the discussion is in terms of molecules diffusing out of an enclosure, similar mathematical formulations may be used to describe processes such as the random migration of plant and animal species to new areas, or the spread of a new gene through a randomly mating population.

Motion of the molecules is assumed to be random in several senses. In order to recognize other kinds of systems that may be treated with the same mathematical formulation, it is important to understand how the different kinds of randomness arise and how they relate to each other. First, the motion of the molecules is random with respect to direction. The direction of an individual molecule is continually changed by collisions with other molecules and with the walls of the vessel. Therefore, even if we knew the direction of travel of a particular molecule at a given instant, it would be a very short time until we would be completely uncertain about it. It follows that for a randomly selected molecule, all directions are equally likely.

Second, the motion is random with respect to velocity. This does not mean that all velocities are equally likely. It means that we can only specify an average velocity, and other characteristics of the distribution of probabilities over the set of possible velocities. The distribution applies equally and independently to every molecule but cannot specify the velocity of a given molecule at any instant of time. This randomness arises because molecules are continually colliding with each other and exchang-

5.5 EXAMPLES OF SIMPLE LINEAR RATE PROCESSES

ing energy with each other. Therefore, even if we knew the velocity of a given molecule at some instant, it would only be a short time (time enough for a few intermolecular collisions) until we were uncertain in the sense just discussed.

Third, the randomness as to direction and velocity results in randomness as to position. That is, after a period of time that depends upon the average velocity and upon the size of the enclosure, there is complete uncertainty as to the position of the molecule; all positions are equally likely. The uncertainty as to position has another important implication; if we select any small region within the enclosure, the expected number of molecules within that region is simply the total number of molecules multiplied by the fraction of the total volume contained in the region (area would be used for discussions of random migration in two dimensions). This number is the same regardless of where the small region is located within the enclosure.

Now suppose we provide the box with a small hole or network of holes, as in a membrane (Figure 5.5). Each molecule can be assigned a probability of escape within some period of time Δt. The same probability applies equally and independently to each molecule. If we let N be the number of molecules remaining in the box, then the process becomes a "pure death" process that can be described by the differential equation for expectations (5.8). If the concentration of molecules is not extremely low (small number per unit volume), the continuous deterministic equation (5.9) is applicable.

It is usually convenient in discussions of diffusion to work in terms of concentrations rather than in terms of total number. The reason is that the expected total number of escapes in a short period of time Δt depends upon the number of molecules in the region immediately adjacent to the membrane. Molecules that are too far away simply cannot get to the membrane within the specified time interval. Under the assumption of randomness, the number per unit volume or concentration, is uniform throughout the enclosure and for a fixed total volume is proportional to the total number of molecules. Concentration has the advantage of being a scale-invariant ratio, which can be used to compare systems of different

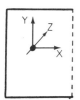

Figure 5.5. Particle moving in a box with holes on one wall.

sizes (see Section 3.7). Equation (5.9) can be converted to an equation for concentration by dividing both sides by the volume. Using k to denote the parameter and C to denote concentration, this leads to

$$\frac{dC}{dt} = -kC \qquad (5.21)$$

For discussion of migration in two dimensions, area would be substituted for volume, and density (number per unit area) would be substituted for concentration. Since the same formulation results when analogous assumptions are made, we use the terms "concentration" and "density" interchangeably and let the context specify whether volume or area is intended.

This model for random migration is a frequently used one, so it pays us to take a moment to examine the applicability of the assumptions.

First, it was assumed that the motion of an individual is completely random as to direction and velocity. This is not true if the space is structured so as to prejudice the motion of the particle, making one direction more probable than another: for example, if there is an electrical field in the space or if the membrane is charged or if the molecules are very heavy (say, macromolecules) and the whole system is placed in a very strong gravitational field, as in an ultracentrifuge.

Homogeneity with respect to time is, as in the case of homogeneity of space, subject to subversion by external forces. For example, the membrane may become clogged.

The assumption that particles behave independently and identically is normally valid if the concentration is not too great. As the concentration goes up, however, various attractive and repulsive forces operate, so that the direction and velocity of motion of the particles are correlated with each other. If the particles are charged, such interactions are of greater importance than if they are neutral.

The assumptions involved in taking the limit as Δt goes to zero and in dropping the expectation symbol prevent the use of equation (5.21) to describe changes over a time period so short that only a few underlying events (escapes) have occurred. These same assumptions prevent use of the equation if the difference ratio does not become essentially constant before Δt shrinks to the time scale on which these underlying events are occurring; such a problem might arise, say, if the membrane ruptured.

Finally, the assumption that all molecules behave randomly and identically can never be completely realized in a system such as this. We must recognize that the particles that are most likely to escape in any interval

5.5 EXAMPLES OF SIMPLE LINEAR RATE PROCESSES

are the ones that are closest to the membrane. Therefore, if a particle is still in the box, it is more likely to be further away from the membrane. This circumstance is at odds with the assumption of randomness, which requires every position to be equally likely. *Equation (5.21) can, therefore, only apply if the rate at which the escapes take place is slow relative to the rate at which the individual particles "re-randomize."* In such cases, equation (5.21) may be a good approximation.

It must be concluded that satisfaction of the randomness assumptions leading to equation (5.21) is not automatic even in molecular systems. When the equation is used to describe migrations in other kinds of systems, such as those mentioned in the introduction to this section, careful attention must be paid to the question of how validly the required assumptions describe the system.

Using equation (5.21), one may proceed to a description of concentration as a function of time,

$$C(t) = C(t_0)e^{-k(t-t_0)} \qquad (5.22)$$

where t_0 is any convenient reference time. The associated concentration curve is qualitatively similar to that in Figure 5.4b.

5.5c. Light Absorption and Beer's Law

In this example, we look at a rate of change with respect to distance, rather than time. The physical setup is shown in Figure 5.6. We shine a beam of monochromatic light (that is, all of the same wave length) on a solution of some chemical compound. The solution is contained in a cell with two completely transparent sides, so that any light not absorbed by the solution passes completely out of the system. The solvent itself is assumed to absorb none of the light. The beam of light is focused so that all of it that is not absorbed passes straight through to the other side. Such conditions can be very closely approximated experimentally, and deviations from the conditions can be corrected for.

We know that the light beam can be considered to be composed of individual particles called photons. Similarly, the solute consists of individuals particles—the molecules. An individual act of absorption takes place when a photon strikes a molecule and is absorbed. The probability of this happening is the product of P(photon strikes a molecule) and P(absorption| striking a molecule).

The molecules in solution are assumed to be stochastically identical and independent. Each increment of length is expected (stochastic expectation)

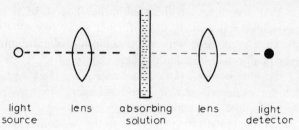

Figure 5.6. Simple representation of apparatus for measuring light absorption.

to have the same number of molecules in it. Each unabsorbed photon passing through has the same probability of being absorbed by any molecule. The photons are assumed to be independent and identical (stochastic sense).

In a given increment of length Δx, a given unabsorbed photon has a certain expectation of being absorbed. The expectations for all photons are identical and independent (analogous to assumption a of Section 5.2); the expectation depends only on the length of the increments Δx and not on where in the cell in the increment begins (analogous to assumption b, Section 5.2), and it goes smoothly to zero as the length Δx is reduced to zero (analogous to assumption c). Now, if a population of M_0 photons enters the cell at the position x_0, and we let $M(x)$ be the number of photons that are able to reach position x, we obtain an equation completely analogous to (5.8); or, if the number of photons is large enough, analogous to equation (5.9),

$$\frac{dM(x)}{dx} = -kM(x) \tag{5.23}$$

The fraction that survive until position x is

$$\frac{M(x)}{M_0} = e^{-k(x - x_0)}$$

If, in the next instant of time, a new population of photons strikes the cell at x_0, the fraction of this new population that reaches position x is exactly the same, providing we assume that the probabilities of absorption do not depend upon time. Such an assumption is satisfied except for special types of situations.

Now suppose we have a steady supply of photons, as we would when a

5.6 NON-HOMOGENEOUS POPULATIONS

beam of light is focused on the cell; that is, we have a constant number of photons entering at x_0 per unit time per unit of surface area of the cell. The number of photons per unit of area is measured by the intensity of the light and is usually denoted by I. From the previous paragraph, it follows that the fraction of intensity remaining at any position x is given by

$$\frac{I(x)}{I_0} = e^{-k(x-x_0)} \qquad (5.24)$$

where I_0 is the incident intensity. Equation (5.24) is known as *Beer's Law*. The formulation may be taken a step further by noting that since the molecules are assumed to behave identically and independently, the parameter k is proportional to the number of molecules per unit volume, that is, to their concentration. The reasoning is similar to that used in the bee-foraging example of Section 5.1. If we take x_0 as zero, then the resulting equation, which is known as the *Lambert–Beer Law*, is

$$\frac{I(x)}{I_0} = e^{-\gamma C x} \qquad (5.25)$$

where γ is called the *absorbancy* and has dimensions of $(\text{length})^{-1} (\text{conc.})^{-1}$. Its value depends upon the identity of the absorbing substance as well as on the wave length of the light. Equation (5.25) finds direct application in analytical biochemistry as a means of determining concentration, in studies relating to vision and photosynthesis, and in the evaluation of light energy input to ecosystems as affected by depth and density of leaf canopy (see subsection 5.8c).

5.6. NONHOMOGENEOUS POPULATIONS

In this section, we consider a process operating on a population that is not homogeneous; that is, the individuals do not behave identically. We assume, however, that the population can be divided into subpopulations and the individuals within each subpopulation behave identically. We continue to assume that the individuals are stochastically independent and that a derivative formulation is acceptable. An example would be the growth of a bacterial culture (or any other biological population) consisting of a mixture of organisms, each with its own birth and death rate parameters. If n_i is the number in the ith group, and κ_i is the corresponding rate parameter, then for each group we have

$$\frac{dn_i}{dt} = \kappa_i n_i \qquad (5.26)$$

If there are r groups, then the total number in the population is $N = n_1 + n_2 + \cdots + n_r$. Because of the assumption of no interaction (i.e., the assumption of independence), the rate of change for N is just the sum of the rates for the n_i, so

$$\begin{aligned}\frac{dN}{dt} &= \frac{dn_1}{dt} + \frac{dn_2}{dt} + \cdots + \frac{dn_r}{dt} \\ &= \kappa_1 n_1 + \kappa_2 n_2 + \cdots + \kappa_r n_r \end{aligned} \qquad (5.27)$$

If we let the initial time be zero and $n_{i,0}$ be the initial value of the ith subpopulation, then from (5.26) we get

$$n_i(t) = n_{i,0} e^{\kappa_i t} \qquad \text{for } i = 1, 2, \ldots, r \qquad (5.28)$$

Therefore,

$$N(t) = n_{1,0} e^{\kappa_1 t} + n_{2,0} e^{\kappa_2 t} + \cdots + n_{r,0} e^{\kappa_r t} \qquad (5.29)$$

An essential difference between this case and the case of a homogeneous population is that the subpopulations change at different rates and possibly in different directions. In the homogeneous population case, the nature of the population remains the same, and the relative rate of change $(1/N)(dN/dt)$ remains constant. In the case of several subpopulations, the make-up of the population is continually changing, and the relative rate of change for the population as a whole is not constant. If all of the rate parameters are negative (all subpopulations decreasing), the population is eventually dominated by the subpopulation that is decreasing most slowly. On the other hand, if any subpopulation has a positive rate parameter, the population will in time be dominated by the subpopulation that is most rapidly increasing.

The mathematical simplicity of the homogeneous case is lost. We cannot simply plot $\ln N$ versus t and get a straight line from which to estimate rate parameters and initial values. However, if all populations are decreasing, a plot of $\ln N$ versus t can yield rough estimates of each of the parameters and of the initial values, providing the magnitudes of the parameters are very different from each other. The procedure for getting such estimates is

5.6 NON-HOMOGENEOUS POPULATIONS

popularly called the "peeling-off" of exponentials. It should be used with caution, and the results regarded with skepticism. It is described because it may provide a convenient first approximation and because an understanding of the method provides some additional insight into the process of change in which the population is involved. Since all subpopulations are taken to be decreasing, we substitute $-k_i$ for κ_i in equations (5.26) to (5.29).

We begin with the case for $r=1$. This is just the single homogeneous population case; a plot of observed values of N versus t would be a straight line, whose slope would be $-k$ and whose intercept would be $\ln n_0$.

Next, assume $r=2$, number the rate constants so that $k_1 > k_2$, and assume that the magnitudes of k_1 and k_2 are *well separated*. In this case, the change in subpopulation 1 is initially faster, and a relatively larger amount of the total change is due to subpopulation 1. If we wait long enough, then group 1 may be essentially depleted, so that most of the remaining change is due to the slower process relating to group 2. A plot of $\ln N$ versus t would then start out with a slope approximating $-k_1$, and would curve around as the relative sizes of the two groups change, so as to approach $-k_2$ in the "tail" (Figure 5.7). If the rates of the two processes are well enough separated, it may be that for sufficiently large t, we get $n_1(t) \approx 0$, while $N(t) \approx n_2(t)$ is still large enough so that a rate of change may be measured. In this case, the plot will approximate, for large t, a straight line with slope $-k_2$. Projecting this straight line back to $t_0 = 0$, the intercept gives an estimate of $n_{2,0}$. Now we have estimates of everything needed to compute $n_2(t)$ for any time t. We then go back to the original equation and "peel off" the estimate of $n_2(t)$. That is, we use

$$N(t) - n_2(t) = n_{1,0} e^{-k_1 t}$$

where $N(t)$ is the measured value and $n_2(t)$ is the calculated value. Plotting $\ln[N(t) - n_2(t)]$ versus t should now give a straight line with slope $-k_1$ and intercept $n_{1,0}$.

To generalize the method to arbitrary (unknown) r, we set up the following steps:

1. Plot $\ln N(t)$ versus t.
2. Estimate the straight line that this plot is approaching.
3. From the line in step 2, determine a rate constant and initial value.
4. For each value of $N(t)$ plotted subtract the value calculated with the parameters from step 3.

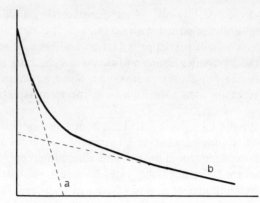

Figure 5.7. Curve starts with slope of line a and bends around to approach line b in the "tail."

5. Plot the logarithm (base e) of the result of step 4 versus t.
6. If the plot is a straight line, then estimate the rate constant and initial value and stop. If it still has the qualitative shape of Figure 5.7, return to step 2.

If everything operates ideally, this yields an estimate of the number of groups together with the rate constant and initial value for each group.

The sources for error in this procedure should be apparent from its description. First, there is the requirement that the rate constants be sufficiently well separated so the processes for each subpopulation may be considered to be effectively terminated one at a time. As a general rule, it is best if the constants are separated by at least a factor of 10. There is usually very little hope of recognizing individual rate constants separated by much less than a factor of three.

Then there is the need to estimate a straight line—not one determined by the data points, but one that the data points are assumed to be approaching! This would be tough enough if the points themselves were error-free. Since this is hardly ever the case, the job gets very tricky indeed.

It is often asserted that even though the procedure may not detect all of the groups (since the k_i may not be sufficiently different), the number of groups detected can be taken as a lower limit to the actual number. This would be so except for the errors involved in estimating an asymptote from data of limited accuracy.

5.7. DIRECT INTERACTIONS; CHEMICAL LAW OF MASS ACTION

In this section, we begin looking at processes in which individuals interact with one another. Chemical systems are used to provide the prototype example because the assumptions that are made are reasonably well satisfied in such systems. Explicit consideration of these assumptions should make it plain why similar expressions are often used as approximate descriptions of interactions in other types of systems and what limitations must be placed on their use. The simplest type of direct interaction is expressed by the law of mass action. This law is developed in subsection 5.7a for interactions involving two individuals and then generalized to interactions involving any number of individuals.

5.7a. Bimolecular Reactions

A reaction between two molecules of different types is expressed by the chemical equation,

$$A + B \rightarrow \text{product}$$

The basic assumptions are that all molecules of A are identical to each other (stochastically), that all molecules of B are identical to each other, and that all the molecules behave independently except for the interactions that are explicitly considered.

Before a molecule of A can react with a molecule of B, the two molecules must come together. We may visualize molecules of A and of B "swimming" around in solution, moving at random. Every once in a while, a molecule of A and one of B encounter each other. Only then is there possibility of reaction. If we pick one particular molecule of A and of B, say a^* and b^*, then the probability that they will react in some time interval Δt is

$$P(a^* \text{ reacts with } b^*) = P(a^* \text{ encounters } b^*)$$
$$\times P(a^* \text{ reacts with } b^* | a^* \text{ encounters } b^*) \quad (5.30)$$

When a^* and b^* encounter each other, they either bounce apart or react with each other. Furthermore, since all molecules of A and all of B are taken to be identical, the probability of the event (a^* reacts with b^*| a^*

encounters b^*) is just the probability that any given encounter between A and B types leads to reaction. This probability is denoted p_r.

Now look at the event (a^* encounters b^* in time Δt). The probability of this event depends on Δt. If a^* and b^* encounter each other, but do not react, they may encounter each other again; if Δt is long enough (say $\Delta t \to \infty$), they continue to encounter each other until eventually they react. However, molecular encounters are discrete physical events, and it is possible to make Δt small enough so that a^* and b^* cannot encounter each other more than once.

In such a time interval, a^* has an expected number of encounters with b^*, say $\lambda = \mathcal{E}[a^* - b^*$ encounters in time $\Delta t]$; this is a number between 0 and 1 because of the way Δt was chosen. But since all molecules of B are alike, a^* has the *same* number of expected encounters with *each* molecule of B. If there are N_B of them, then the expected number of encounters that a^* has with B molecules is λN_B. Furthermore, each molecule of A has the same expected number of encounters, so the total expected number of $A - B$ encounters during the time interval Δt is $\lambda N_A N_B$.

Each of these encounters has the same probability of leading to reaction. If we let p_r denote this probability, then the expected number of reaction events between a^* and b^* in time Δt is λp_r, and the expected number of reaction events between all $A - B$ molecules is $\lambda p_r N_A N_B$.

Each reaction event decreases the number of A molecules by one. Therefore, the expected change in N_A during Δt is

$$\mathcal{E}\left[N_A(t+\Delta t)\right] - \mathcal{E}\left[N_A(t)\right] = -\lambda p_r N_A(t) N_B(t) \tag{5.31}$$

The next step is to make an assumption analogous to that of assumption b of Section 5.1—that the probabilities go smoothly and continuously to zero as Δt approaches zero. The result is that if Δt is small enough, λp_r can be taken to be approximately proportional to Δt. Letting $\lambda p_r \cong \lambda' \Delta t$, equation (5.31) becomes

$$\mathcal{E}\left[N_A(t+\Delta t)\right] - \mathcal{E}\left[N_A(t)\right] = -\lambda' N_A(t) N_B(t) \cdot \Delta t + o(\Delta t) \tag{5.32}$$

As in Section 5.2, if the values of N_A and N_B are uncertain, this can be written

$$\mathcal{E}\left[N_A(t+\Delta t)\right] - \mathcal{E}\left[N_A(t)\right] = -\lambda' \mathcal{E}\left[N_A(t) \cdot N_B(t)\right] \cdot \Delta t + o(\Delta t) \tag{5.33}$$

5.7 DIRECT INTERACTIONS; CHEMICAL LAW OF MASS ACTION

Dividing by Δt and taking limits as $\Delta t \to 0$,

$$\frac{d\mathcal{E}(N_A)}{dt} = -\lambda' \mathcal{E}[N_A(t) \cdot N_B(t)] \tag{5.34}$$

If conditions are met for dropping the expectation signs,

$$\frac{dN_A(t)}{dt} = -\lambda' N_A(t) N_B(t) \tag{5.35}$$

The term $\lambda' N_A \cdot N_B$ is the *rate of reaction*, or the rate of the process viewed as a continuous deterministic process. It is the rate at which both N_A and N_B decrease.[5]

Finally, we note that the probability of an $A - B$ encounter is inversely proportional to the volume in which the molecules move. Therefore, we let

$C_A = N_A/V$, $C_B = N_B/V$, and $k = \lambda' \cdot V$. Equation (5.35) becomes

$$\frac{dC_A}{dt} = \frac{dC_B}{dt} = -kC_A C_B \tag{5.36}$$

5.7b. Reactions Between Two Molecules of the Same Type

Here we consider a reaction of the type,

$$2A \to \text{product}$$

The development is just the same as in subsection 5.7a except that a given molecule a^* has only $N_A - 1$ other A molecules to encounter. This is true for each of the N_A molecules. The expected number of reactions in a small time Δt is then $\lambda' N_A(N_A - 1)\Delta t$. The rate of reaction (expected number of reaction events per unit time) may be expressed

$$\text{rate of reaction} = \lambda' N_A (N_A - 1)$$

---

[5]In Section 5.2, it is noted that the rate given by the deterministic differential equation exactly agreed with the rate given for the expectation of the stochastic model. This seems to be the case here also, but it is not necessarily so. The difficulty arises because the expectation of a product is not equal to the product of the expectations unless the variables are uncorrelated, which is not true in this case. This point is discussed in some detail by Smeach and Gold (1975).

In this case, each elementary reaction event leads to the disappearance of *two* molecules of A. The rate of reaction is, therefore, only half the rate of disappearance of A molecules. In place of equation (5.35), we get

$$\frac{1}{2}\frac{dN_A}{dt} = -\lambda' N_A (N_A - 1) \tag{5.37}$$

Of course, if N_A is large enough so that it is approximately valid to use derivatives, then N_A is not significantly different from $N_A - 1$. Therefore, equation (5.37) can be written

$$\frac{1}{2}\frac{dN_A}{dt} = -\lambda' N_A^2 \tag{5.38}$$

or in terms of concentration,

$$\frac{1}{2}\frac{dC_A}{dt} = -kC_A^2 \tag{5.39}$$

5.7c. The General Mass Action Law

Now suppose the reaction involves some fixed number a of A molecules, a fixed number b of B molecules, a fixed number c of C molecules, and so on,

$$aA + bB + cC + \cdots \rightarrow \text{products}$$

In the chemical literature, the fixed numbers, a, b, and c are called the *stoichiometric coefficients*. The law of mass action states that the rate of reaction is proportional to the product of the concentrations of the reactants, each raised to the power of its own stoichiometric coefficient,

$$\text{rate of reaction} = kC_A^a C_B^b C_C^c \cdots \tag{5.40}$$

Comparison with equations (5.36) and (5.39) should make it clear how such a result is arrived at in terms of the assumed behavior of the individual molecules.

One of the most important assumptions, because it is most often violated, is the assumption that the molecules behave independently except for the specific interaction considered. Other interactions begin building up

5.8 NON-HOMOGENEITY OF THE INDEPENDENT VARIABLE

even at moderate concentrations, so that equation (5.40) does not strictly apply except in dilute systems.[6]

It is a simple matter to relate expression (5.40) to the rate of disappearance of the individual reactant species. We just have to keep in mind that each reaction event leads to the disappearance of a molecules of A, b molecules of B, and so on, so that

$$\frac{1}{a}\frac{dC_A}{dt} = \frac{1}{b}\frac{dC_B}{dt} = \frac{1}{c}\frac{dC_C}{dt} = \cdots$$

$$= \text{rate of reaction}$$

$$= -kC_A^a C_B^b C_C^c \cdots \qquad (5.41)$$

5.7d. The Assumptions and More General Applications

Once again recall the package of assumptions that goes into the kind of expressions we have been considering. All individuals (molecules in this case) are assumed to behave independently except that certain events require simultaneous participation by more than one individual. All individuals of a given species are taken to be identical (stochastic sense). Both space and time are taken to be unstructured. Added to these are the assumptions relating to the use of derivatives of variables that can only have integer values.

In a more general view, a *direct interaction* event may be regarded as an event that requires the presence of two or more individuals at the same time in the same place. In the following chapters and in Appendix A, expressions comparable to those developed in this section are used to represent direct interactions between individuals that are as tangible as molecules and biological organisms or as abstract as increments of energy or informational signal. The applicability of the mathematical formulation depends only upon the acceptability of the assumptions.

5.8. NONHOMOGENEITY WITH RESPECT TO THE INDEPENDENT VARIABLE

All of the processes thus far considered in this chapter have been assumed to be homogeneous with respect to the independent variable. What

---

[6]To take this fact into account, chemists have built up a marvelous array of esoteric terminology (activity, virials, fugacity, etc.).

happens in an interval of time (or of distance, as in the light absorption example of subsection 5.5c) depends only on the state of the system—as expressed by the values of the state variables—at the beginning of the interval. In such a description, the values of the parameters do not change. The underlying events have no explicit dependence on when (or where) the interval is chosen, and, consequently, the derivatives have no explicit dependence on the independent variable; the independent variable does not appear on the right-hand side of any of the differential equations. However, this type of idealization is rarely satisfied in a biological system.

Explicit dependence on the value of the independent variable can often be attributed to incomplete specification of the system state or failure to describe the state of the environment with which the system is interacting. In Chapter 2, for example, it was pointed out that the values of the parameters for the public address system could be sensitive to environmental conditions as well as to the accumulated wear. Another example is the process of diffusion of metabolic substances through cell membranes, described in subsection 5.5b. In this case, the parameter describes the permeability of the membrane, which may change with the age of the cell. In the bee-foraging example, it seems reasonable that the parameter would depend upon time of day[3] and perhaps the length of time since the start of the foraging activity.

The behavior of systems described by differential equations that have time-dependent parameters depends upon the way in which the parameter changes. Determination of how the parameters of a system change with time (or more generally, with the value of the independent variable) may be the specific object of a research project.

A special problem arises in the study of systems in which the underlying events are synchronized. The use of *difference equations* to describe such systems is briefly discussed in subsection 5.8a.

When the independent variable is other than time, it often happens that the effects of a new increment (i.e., interval) of the independent variable depends on the accumulated but unspecified effects of past increments. For example, radiation damage caused by a new increment of radiation *increases* as the total amount of radiation is increased, providing the total dose is not large enough to be saturating. The amount of increased growth produced by a certain increase in nutrient intake tends to *decrease* as the total level of nutrient intake is increased. Explanatory models for such changes in sensitivity usually require consideration of interactions between several simple processes and a fuller specification of the system state. This is reserved for the next chapter. In this section (subsections 5.8b and 5.8c),

5.8 NON-HOMOGENEITY OF THE INDEPENDENT VARIABLE

we look at how these effects may be qualitatively described in the spirit of subsection 2.7e.

5.8a. Synchronously Growing Populations

Many types of biological populations are synchronized in time. Bacterial cultures may be grown in which all cells are simultaneously at corresponding points in time (in the sense of subsection 3.6d) with respect to cell division, although the population gradually "rerandomizes" itself after several generations. Many natural populations are synchronized by environmental conditions (including temperature, moisture, and day length), especially in temperature climates. Agricultural systems are synchronized by a combination of environmental conditions and managerial practices; at a given time of year, a field is always in the same state to within the farmer's ability to make it so.

Insect populations in temperate climates provide striking examples of naturally synchronized systems. We may specifically cite the life cycle of the gray larch budmoth (*Zeiraphera griseana*) (summarized by Clark et al., 1967): eggs hatch from late May to mid-June; five instars (stages of growth) are passed through in about 50 days; adults emerge about 30 days later; adults live for about 35 days during which the females lay their eggs. The nature of the events within the population is changing synchronously. This structure with respect to time would have to be accounted for in a mathematical description of the population growth. Compartmental decompositions of the life cycle are often used for this purpose (see subsection 2.4b and Section 6.3); each stage of development is taken as a compartment, and the equations are formulated to describe the rate of transfer or the probability of transfer from one compartment to the next.

Often, however, it is desired to describe the year-to-year progress of the population without regard to the fine structure of the yearly life cycle. In doing so, one might let N be the number in a particular growth stage, say number of first instar larvae. The population at a given time t would be denoted $N(t)$, and the population one year later would be $N(t+\Delta t)$, where Δt has the value one year. The equations would then express $N(t+\Delta t)$ as a function of $N(t)$ together with whatever other factors must be taken into account. This type of description involves averaging over all the events that have occurred during the year and requires a large enough population for averages to be meaningful. In such a description, we do not allow Δt to approach zero so as to produce a differential equation. The description is in terms of finite differences over finite periods of time.

Although difference equations are somewhat clumsier to handle than differential equations[7], they are of particular use whenever there is a value of Δt that is a natural characteristic of the system. In any system that exhibits cyclic behavior, the cycle period would be a natural choice for Δt.

If we now change perspective and ask for a description of change on a time scale that is long compared with the time for a single cycle, it may be possible to use a differential equation formulation. The derivative in this case would be a rate of change averaged over a number of cycles. The condition for its use would be that the "average rate of change" remain constant over a sufficient number of cycles for the use of an average to be meaningful, that is, that the ratio $[N(t+\Delta t) - N(t)]/\Delta t$ change slowly compared to the time per cycle. In the example of the insect population, this might lead to an expression for dN/dt, where N would be the number of individuals in a particular growth stage at a particular time of year, say, the number of first instar larvae in the second week of June. In a comparable sense, one may, for example, speak of the weight change per day for an animal that exhibits diurnal weight periodicity or discuss the rate of change of the annual agricultural output.

5.8b. Increasing Sensitivity

In this case, each succeeding increment of the independent variable x produces a greater effect on the dependent variable y. This is reflected by the positive slope of a plot of dy/dx versus x. The question that concerns us is how to relate the mathematical description of such a curve to the kinds of mechanisms that might lead to such behavior.

By way of example, consider the chemical reaction,

$$nA + B \rightarrow \text{product}$$

We ask how the rate of disappearance of B depends upon the concentration of A. Letting C_A and C_B be the concentrations of A and B and v be the rate of disappearance, the law of mass action gives the relation

$$v = kC_A^n C_B$$

The dependence of v on C_A is

$$\frac{dv}{dC_A} = nkC_B C_A^{n-1} \tag{5.42}$$

---

[7] An introduction to the mathematics of finite difference equations may be found in Grossman and Turner (1974).

5.8 NON-HOMOGENEITY OF THE INDEPENDENT VARIABLE

In the process, we can consider that the destruction of B is "caused by" substance A, and that n molecules (individual units in a more general context) of A must cooperate to bring about a single destruction event. A plot of dv/dC_A versus C_A for a fixed value of C_B is a curve whose slope increases in proportion to the $(n-1)$st power of C_A. If $n=1$, then there is no cooperativity; dv/dC_A would show no dependence on the value of the independent variable C_A.

A specific example of increasing sensitivity may be taken from the target theory of biological radiation damage. If multiple "hits," say n of them, are needed to cause irreversible damage, then for very small radiation doses, the extent of the damage is proportional to the nth power of the dose level. Therefore, for *small doses* (see Pollard, 1969, for discussion),

$$\frac{d(\text{damage})}{d(\text{dose})} = \text{const.} \times (\text{dose})^{n-1} \tag{5.43}$$

Equations (5.42) and (5.43) are both of the form

$$\frac{dy}{dx} = k_1 x^{k_2} \tag{5.44}$$

where k_1 and k_2 are parameters. This is one of the commonly used mathematical forms for the description of increasing sensitivity to the independent variable. A plot of dy/dx versus x gives a curve as in Figure 5.8, where the curvature depends on k_2. If $k_2=1$, then the curve is simply a

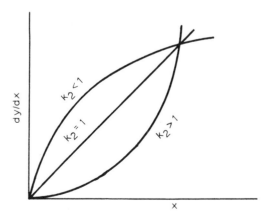

Figure 5.8. Plots of equation (5.44) for different values of k_2. Note that if all curves have the same value for k_1, they meet at $x=1$.

straight line. If $0 < k_2 < 1$ (as might arise if the cooperativity is only partial), then the slope is still positive, but the curvature is downward. In the limiting case of $k_2 = 0$, the value of the derivative dy/dx does not depend on x. For $k_2 < 0$, the situation becomes one of *decreasing* sensitivity, which is discussed in subsection 5.8c.

The behavior of y as a function of x may be obtained by directly integrating equation (5.44)—providing that all the change in y is associated with the change in x. The result is

$$y(x) = y_0 + \frac{k_1}{k_2+1} x^{k_2+1}$$

5.8c. Decreasing Sensitivity (Diminishing Returns)

In this case, the increments of the independent variable exert an effect whose magnitude decreases as the total amount increases. The effect can often be interpreted as a saturation effect. For example, if y is the growth rate of an animal or a plant and x is the concentration of some limiting nutrient in the diet (or in the soil), then dy/dx would express the dependence of growth rate on the availability of the nutrient (see, for example, Parks, 1972). Response to added nutrient is large, if the total amount available is small. If plenty is available, a supplement has little or no effect.

Other examples might be dependence of the rate of plant growth on light intensity, dependence of the rate of an enzyme reaction on the availability of substrate, disease remission versus drug concentration. The reader should have no trouble making a list of examples from his own field of biology.

Two alternative mathematical expressions that might be used to describe this type of relation between variables y and x are

$$\frac{dy}{dx} = k_1 e^{-k_2 x} \tag{5.45}$$

and

$$\frac{dy}{dx} = k_1 (x - \theta)^{-k_2} \tag{5.46}$$

where $k_2 > 0$ and (usually) $k_1 > 0$. The parameter θ in (5.4b) is needed to prevent the description from being unrealistically unstable as $x \to 0$. The parameter θ plays the role of a reference point value.

If we assume k_1 to be positive, then for both equations, dy/dx is positive but gets progressively smaller as x gets large and approaches zero asymptotically as $x \to \infty$ (Figure 5.9). It immediately follows that a plot of y versus

5.8 NON-HOMOGENEITY OF THE INDEPENDENT VARIABLE

x is a curve with a slope that is positive but becomes less positive as x gets larger and larger.

If the system is following equation (5.45), then the curve of y versus x will flatten out and asymptotically approach a finite constant value (Figure 5.9b). Integration of (5.45) gives

$$y(x) = \text{const.} - \frac{k_1}{k_2} e^{-k_2 x} \qquad (5.47)$$

By taking $x = 0$, we find that the constant is $y(0) + k_1/k_2$. As x becomes larger, the exponential term becomes smaller and becomes zero as x becomes infinite, so that y approaches the asymptotic value of $y(0) + k_1/k_2$. Equations of the type of (5.47) are often called laws of diminishing returns.

If the system is described by equation (5.46), it can be shown that the curve of y versus x flattens out and approaches a finite constant value only if $k_2 > 1$. One might say that y has a finite limit only if dy/dx gets small faster than x gets large.

An example of diminishing returns is provided by the application of Beer's Law (equation 5.24) to the calculation of the relation between total light absorption and depth of leaf canopy in an ecosystem. From equation (5.24), the total intensity absorbed by a canopy of uniform density and of depth x is

$$I_0 - I(x) = I_0 - I_0 e^{-kx}$$

Dependence of this quantity on canopy depth is

$$\frac{d[I_0 - I(x)]}{dx} = kI_0 e^{-kx}$$

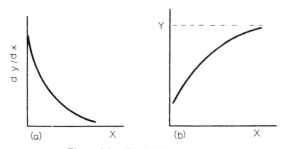

Figure 5.9. Diminishing returns.

This is precisely of the form of equation (5.45). It follows that additional investment in leaf canopy depth leads to a diminishing return of total light absorption. Once a certain canopy depth has been reached, the metabolic energy cost to add more would exceed the light energy return.

5.9. SUMMARIZING REMARKS

In the formulation of an equation to describe a dynamic process, one may begin by constructing a mental image or concept of the elementary events that comprise the overall process. The construction of such a mental image is based on accumulated biological and physical knowledge of the system; it depends heavily on the "biological intuition" of the modeler. In this chapter, we have focused on the derivation of the mathematical description of a dynamic process from knowledge and intuition concerning the laws governing the elementary underlying events. Of course, we have not examined all types of processes; indeed, we have barely scratched the surface. We *have* considered some of the more commonly encountered types of processes, and in doing so have tried to illustrate the general principles of such a derivation.

When the operation of a system involves dynamic processes, the formulation of the mathematical model must recognize the time lags discussed in the introduction to this chapter. It is not unusual to explicitly display the lags as part of the signal–flow graph, as shown in Figure 5.10. For a finite difference expression, both $y(t)$ and $y(t+\Delta t)$ are shown in the graph as separate variables. For a derivative expression, y and $\dot{y}$ (i.e., dy/dt) are shown as separate variables. For the signed digraph, using the symbolism of Section 2.6, it is always true that $\dot{y} \Rightarrow y$.

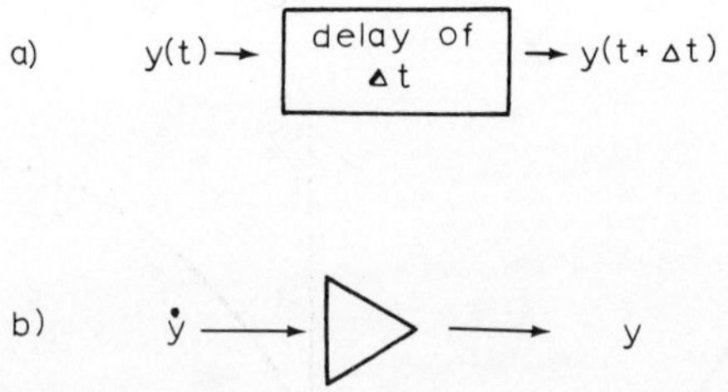

Figure 5.10. Display of time delays in a dynamic process: *a*) In a finite difference model, the delay of Δt is usually shown as a box; *b*) in a derivative model, integration is usually shown by a triangle.

CHAPTER
SIX
INTERACTING DYNAMIC PROCESSES

The centipede was happy quite
Until a toad in fun
Said, "Pray, which leg goes after which?"
That worked her mind to such a pitch
She lay distracted in a ditch
Considering how to run.
Pinafore Poems, 1871
Mrs. Edward Craster

In Chapter 5 we examined the development of laws governing the operation of individual dynamic processes involving the change of a single variable. Most often these processes either lead to an unlimited increase in the variable or else the variable goes to zero, terminating the process.

In this chapter, we begin to look at the simultaneous operation of several processes. The central theme of this chapter is that the counterbalancing effect of opposing processes provides the opportunity for the stabilization of state variables at finite, nonzero values.

In studying the simultaneous operation of several processes, it is usually necessary to consider changes in several variables, so that the state spaces are multidimensional. The mathematical manipulation of models for such systems usually involves the formalism matrix algebra. In this book, the emphasis is on the model formulation rather than on manipulation, so we are able to avoid the use of matrix algebra. However, since state spaces for biological systems are nearly always multidimensional, a biologist is well advised to have a basic knowledge of matrix algebra. Elementary treatments may be found in the references of subsection 1.8b.

Descriptions of multivariable systems make use of various kinds of simplifications to reduce the number of variables that must be considered at one time. Some of the more commonly used are:

a. **Construction of a Separate Description for the Progress of Each Variable.** This is complicated by the fact that the progress of any one variable is usually influenced by all other variables.

b. **Time Scale Separation of Variables.** Two processes may be thought of as operating on different time scales if their relaxation times (subsection 5.4b) are so different that one process goes essentially to completion while the other has made negligible progress. This allows each process to be treated separately. This type of simplification was made use of in the example of diffusion across a barrier (subsection 5.5b), where it was assumed that the process of rerandomization is so fast that it does not have to be explicitly considered. It is also the basis of the "peeling off" of exponentials, described in Section 5.6.

c. **Combining Several Variables into One.** A frequently used simplification is to define a single new variable that is a function of several state variables (i.e., their sum or difference). The description is then constructed in terms of this single variable. This type of simplification is often compelled by inability to experimentally measure each variable separately.

6.1. ISOLATED SYSTEMS AND EQUILIBRIUM

An isolated system is defined as one that exchanges neither matter nor energy with its environment; it is isolated from the environment. It is the simplest kind of system within which one may study the balancing of opposing processes; we, therefore, begin by looking at such a system, even though it must be realized that no biological system is isolated in this sense.

The example that is considered is an extension of that of subsection 5.5b and is shown in Figure 6.1. The system has two compartments and two transfer processes: i) transfer from I to II; ii) transfer from II to I. Each of the processes operates by the rules developed in subsection 5.5b, which you should take a moment to review. The state variables are the amounts or concentrations (densities) in the two compartments. The overall system is isolated; transfers may take place from one compartment to the other, but nothing enters or leaves the overall system.

6.1 ISOLATED SYSTEMS AND EQUILIBRIUM

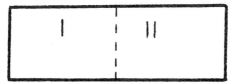

Figure 6.1. A simple two-compartment system.

6.1a. Formulation of the Model

To the ground rules of subsection 5.5b, we add the following two auxiliary assumptions:

i. If an individual hits the barrier, it has the same probability of passing through, regardless of which side it hits from. In a more general context, the assumption would be that the individual members of the population cross the barrier with equal ease from either direction.
ii. Migrations in the two directions are independent in the following sense: Once an individual hits the barrier, the probability that it passes through does not depend on how many individuals are traveling in the opposite direction.

It is easy to see that if the concentrations get high enough, we are going to have to reckon with the probability of collisions, so that ii won't hold.

As always, failure to explicitly examine the applicability of the underlying assumptions may lead (and often does) to inaccurate models. In practice, the assumptions used in this section apply, at least as a reasonable first approximation, for a wide variety of systems.

Accepting these assumptions, we find from subsection 5.5b that the rate of movement from I to II is proportional to the number of individuals in compartment I, which we call X_1, and inversely proportional to the volume of compartment I,

$$\text{rate (I} \rightarrow \text{II)} = k \frac{X_1}{V_1} \qquad (6.1a)$$

Similarly,

$$\text{rate (II} \rightarrow \text{I)} = k \frac{X_2}{V_2} \qquad (6.1b)$$

The same parameter is used in (6.1a) and (6.1b) because of auxiliary assumption i.

Now, to determine how the amount in each compartment changes, we note that for each compartment,

$$\text{rate of change} = \text{rate in} - \text{rate out}$$

Applying this to compartment I,

$$\frac{dX_1}{dt} = k\frac{X_2}{V_2} - k\frac{X_1}{V_1} \tag{6.2}$$

If we let $C = X/V$ be the concentration (or density), then this becomes

$$\frac{dX_1}{dt} = k(C_2 - C_1) \tag{6.3}$$

Equations (6.2) and (6.3) express the interaction between the compartments; the rate of change of X_1 is influenced by the density in compartment II. The interaction is *indirect*; the "population" X_2 influences the "population" X_1 by contributing to it but does not directly influence the behavior of any individual members of population X_1.

If we let $X_T = X_1 + X_2$, then because the system is isolated,

$$\frac{dX_T}{dt} = \frac{dX_1}{dt} + \frac{dX_2}{dt}$$
$$= 0$$

If we let ΔC be the density difference,

$$\Delta C = C_2 - C_1$$

this leads to

$$\frac{dX_1}{dt} = k\Delta C \tag{6.4a}$$

$$\frac{dX_2}{dt} = -k\Delta C \tag{6.4b}$$

The arrangement of signs in (6.4) coincides with the requirement that the

6.1 ISOLATED SYSTEMS AND EQUILIBRIUM

net flow is from the compartment of higher density to that of lower density. That is, if $C_2 > C_1$ then ΔC and dX_1/dt are positive and dX_2/dt is negative. The signs are reversed if $C_2 < C_1$.

Often it is the change in density (concentration $= X/V$), which is of interest. Frequently, it is the only thing that can be directly measured. To convert equations (6.4) into statements about dC_1/dt and dC_2/dt, we first note that if the volumes are kept constant, then we have the general relation, $(1/V)dX/dt = d(X/V)/dt$. Next, we let $k_1 = k/V_1$. This is a ratio of constants, and so is a constant (what are its dimensions?). Similarly, $k_2 = k/V_2$. Now, dividing (6.4a) by V_1 and (6.4b) by V_2, we get, respectively,

$$\frac{dC_1}{dt} = k_1 \Delta C \qquad (6.5a)$$

$$\frac{dC_2}{dt} = -k_2 \Delta C \qquad (6.5b)$$

It is often convenient to replace the variables C_1 and C_2 by the single variable ΔC. Since $\Delta C = C_2 - C_1$, we have

$$\frac{d(\Delta C)}{dt} = \frac{dC_2}{dt} - \frac{dC_1}{dt}$$

or

$$\frac{d(\Delta C)}{dt} = -(k_2 + k_1)\Delta C \qquad (6.6)$$

Here we have a linear differential equation, with ΔC as the variable and $-(k_1 + k_2)$ as the parameter. As a function of time, ΔC would be

$$\Delta C(t) = \Delta C_0 e^{-(k_1 + k_2)t} \qquad (6.7)$$

The curve would be similar to that of Figure 5.4b.

6.1b. Equilibration Between Compartments

Using equation (6.6), we investigate the time course of the state variables X_1 and X_2. Some initial condition is needed and the most convenient is to assume that at time $t=0$, all material is in compartment I, so that

$$X_1(0) = X_T$$
$$X_2(0) = 0$$

With these initial values, equation (6.7) can be rewritten as

$$\frac{X_2}{V_2} - \frac{X_1}{V_1} = -\frac{X_T}{V_1}e^{-(k_1+k_2)t} \qquad (6.8)$$

Recall that k_1 and k_2 were taken as inversely proportional to V_1 and V_2. Since $X_T = X_1(t) + X_2(t)$ at all times, one of these three can be eliminated and expressed in terms of the other two. It would not be convenient to eliminate X_T, since it is a constant (because the system is closed) and may be known. Elimination of X_2 would give an equation for compartment I. Using $X_2(t) = X_T - X_1(t)$, we get

$$\frac{X_T - X_1(t)}{V_2} - \frac{X_1(t)}{V_1} = -\frac{X_T}{V_1}e^{-(k_1+k_2)t} \qquad (6.9)$$

A little rearranging gives

$$\frac{X_T}{V_2} - \left(\frac{1}{V_2} + \frac{1}{V_1}\right)X_1(t) = -\frac{X_T}{V_1}e^{-(k_1+k_2)t}$$

$$\left(\frac{1}{V_2} + \frac{1}{V_1}\right)X_1(t) = X_T\left(\frac{1}{V_2} + \frac{1}{V_1}e^{-(k_1+k_2)t}\right)$$

Condensing the coefficient of $X_1(t)$ into one fraction and multiplying both sides by the resulting denominator, we wind up with

$$X_1(t) = X_T\left(\frac{V_1}{V_1+V_2} + \frac{V_2}{V_1+V_2}e^{-(k_1+k_2)t}\right) \qquad (6.10)$$

The next thing to do is to see what the time curve for $X_1(t)$ *looks* like. The details, of course, depend upon the values of k_1, k_2, V_1, and V_2. However, we can find the general shape of the curve without the use of specific values. Examination of the shape of this curve is one of the opportunities for comparison between intuition and the mathematics. As

6.1 ISOLATED SYSTEMS AND EQUILIBRIUM

pointed out in Chapter 1, disagreement should always be regarded as a danger signal.

The easiest places to start are at the limits, where $t \to 0$ and $t \to \infty$. For $t = 0$, the exponential term is equal to one, so the whole parenthesized expression is equal to one. The equation reduces to

$$X_1(0) = X_T$$

This was one of the starting assumptions.

As $t \to \infty$, the exponential term goes to zero, and we get

$$\lim_{t \to \infty} X_1(t) = X_T \frac{V_1}{V_1 + V_2}$$

This relation tells us that as time goes on, the amount remaining in compartment 1 approaches a constant value. The compartment retains an amount that depends upon its fraction of the total volume. To put it another way, the total starting amount distributes itself between the two compartments in proportion to their relative volumes. The limiting concentrations or densities as $t \to \infty$ would be

$$C_1(\infty) = \frac{X_T}{V_1} \cdot \frac{V_1}{V_1 + V_2}$$

$$= \frac{X_T}{V_1 + V_2}$$

$$C_2(\infty) = \frac{X_T}{V_2} \cdot \frac{V_2}{V_1 + V_2}$$

$$= \frac{X_T}{V_1 + V_2}$$

The conclusion is that as time goes by, the system approaches an equilibrium state in which the *concentrations* of the two compartments are equal. When the concentrations are equal, then equation (6.6) says that no further change takes place. They are held equal by the balance between the two opposing processes. The *amount* (value of X_1 or X_2) needed for a given concentration depends on the volume in which that amount is to be distributed.

We next ask how fast X_1 gets from its initial value of X_T to its final value of $X_T V_1/(V_1 + V_2)$. It is easier to answer this question if equation (6.10) is rewritten so that the difference between X_1 and its final value is on the left

$$X_1(t) - X_T \frac{V_1}{V_1 + V_2} = X_T \frac{V_2}{V_1 + V_2} e^{-(k_1 + k_2)t} \qquad (6.11)$$

This *difference* starts out to be $X_T[V_2/(V_1 + V_2)]$ at time $t = 0$ and decreases to zero exponentially as $t \to \infty$. The resulting curve is shown in Figure 6.2.

Notice, by the way, that the system of one leaky compartment discussed in subsection 5.5b is just a special case of this system. That is, if compartment II is the whole universe, then V_2 is essentially infinite so that the fraction $V_1/(V_1 + V_2)$ is essentially zero.

To study the time course of X_2, we can eliminate X_1 from (6.8). After a little rearrangement, we wind up with

$$X_2(t) = X_T \frac{V_2}{V_1 + V_2} (1 - e^{-(k_1 + k_2)t}) \qquad (6.12)$$

This equation says that $X_2(0) = 0$ and that X_2 approaches $X_T V_2/(V_1 + V_2)$ as $t \to \infty$. Both statements agree with the conclusions already reached.

The difference between the value of X_2 at any given time, and the asymptotic value of $X_T V_2/(V_1 + V_2)$ is

$$X_2(t) - X_T \frac{V_2}{V_1 + V_2} = -X_T \frac{V_2}{V_1 + V_2} e^{-(k_1 + k_2)t}$$

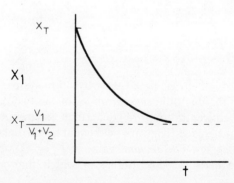

Figure 6.2. Time curve showing decrease in X_1, according to equation (6.11), starting from $X_1 = X_T$.

6.1 ISOLATED SYSTEMS AND EQUILIBRIUM

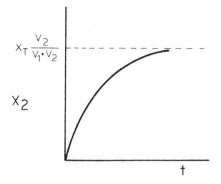

Figure 6.3. Time curve showing increase in X_2 according to equation (6.12) starting with $X_2 = 0$.

This difference is a negative quantity whose magnitude is decreasing exponentially, so the shape of the curve is like that of Figure 5.4d. The difference starts out at $-X_T V_2/(V_1 + V_2)$ and approaches zero, whereas X_2 starts out at zero and approaches $X_T V_2/(V_1 + V_2)$. The curve has the appearance shown in Figure 6.3.

Although the discussion of this section has been in the context of a specific type of system, the result that the system asymptotically approaches equilibrium is a general result that applies to all systems that *do not exchange with the environment*. This is a fundamental principle of thermodynamics. Unfortunately, failure to remember that this principle applies directly only to isolated systems (i.e., systems that do not exchange energy or matter with the environment) has led to the writing of a great deal of nonsense about the violation of thermodynamic principles by living systems.

6.1c. The State-Space Diagram

The state-space diagram portrays the relation between the state variables as they both change. For the system being examined, the relation between the two variables X_1 and X_2 is easily obtained from

$$X_1(t) = X_T - X_2(t) \tag{6.13}$$

That is, the relation is a straight line with slope -1 and intercept X_T. The diagram is shown in Figure 6.4a. The significance of the diagram is that as the system evolves in time, the state must always be a point on this line. In deriving equations (6.10) and (6.12), it was assumed that the initial state

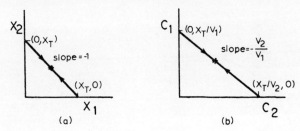

Figure 6.4. State-space representations for isolated two-compartment system. In (a), the state variables are amounts; in (b), they are concentrations.

was $(X_1(0), X_2(0)) = (X_T, 0)$. More generally, the initial state could be any state *on this line* (assuming X_T is fixed in advance).

The star on the line represents the equilibrium state, with coordinates $X_1 = X_T V_1/(V_1 + V_2)$ and $X_2 = X_T V_2/(V_1 + V_2)$. The arrows indicate that as time moves forward, the system moves toward the equilibrium state, regardless of which side it starts from. The diagram does not, however, indicate how fast it moves. For this information, we must turn to equations of the type of (6.10) and (6.12) and diagrams such as Figures 6.2 and 6.3. These indicate that the motion of the system in state space slows down as it gets closer to the equilibrium state. The result is that whereas the system is always headed toward the equilibrium state, it never actually gets there except as time becomes infinite. As a practical matter, however, we might declare it to *be* there when the distance from equilibrium is no longer measurable.

To obtain a state-space diagram using the variables C_1 and C_2, we can divide equation (6.13) by V_1 to get

$$\frac{X_1}{V_1} = C_1 = \frac{X_T}{V_1} - \frac{X_2}{V_1}$$

Then, multiplying and dividing the last term by V_2, we get

$$C_1(t) = \frac{X_T}{V_1} - \frac{V_2}{V_1} C_2(t) \qquad (6.14)$$

This is a straight line (Figure 6.4b) with intercept X_T/V_1 and a slope of $-V_2/V_1$. Its intercept on the C_1-axis (that is, the value of C_1 when C_2 is zero) is X_T/V_2, and the equilibrium point is the point at which $C_1 = C_2 = X_T/(V_1 + V_2)$.

6.1 ISOLATED SYSTEMS AND EQUILIBRIUM

6.1d. Chasing a Moving End Point

As a prelude to consideration of open (nonisolated) systems, we examine what might happen if the equilibrium point that is being approached is changed. That is, suppose that after equilibrium is reached, a second amount X_T' is suddenly introduced into compartment I. The whole process begins over, and both compartments begin to approach the new end point determined by $X_T + X_T'$. The overall time curve might be as shown in Figure 6.5.

Alternatively, suppose that a certain amount X_R is suddenly removed, say from compartment II after equilibrium has been essentially attained. In this case, the end points are changed to $(X_T - X_R)[V_i/(V_1 + V_2)]$, where $i = 1$ or 2. The time curve might be as shown in Figure 6.6.

A more general situation is one in which the end point is being continually changed. If the end point change is very slow compared with the speed with which the system responds (as measured by the values of k_1 and k_2), then we may still have the situation in which the *distance* between

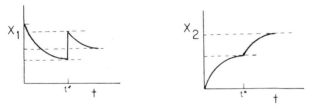

Figure 6.5. Time curves for amounts in two compartments. An amount X_T is introduced into compartment I at time $t = 0$, and a further amount is introduced to compartment I at time $t = t^*$, after equilibration.

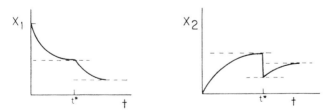

Figure 6.6. Time curves for amounts in two compartments. An amount X_T is introduced into compartment I at time $t = 0$ and a quantity is removed from compartment II at time $t = t^*$, after equilibration.

the value for a given compartment and the asymptotic value (which is now changing) is decreasing exponentially. Graphically, we might represent it as in Figure 6.7.

However, if the end point is changed rapidly, we may have the following kind of situation: The system is always heading in the direction of the end point, but the end point keeps changing direction and moving too rapidly for the system to catch up.

It is especially instructive to look at the situation in which the end point keeps changing in a cyclic fashion. If the response of the system is very fast

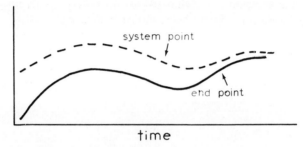

Figure 6.7. System catching up with moving endpoint.

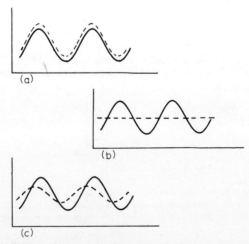

Figure 6.8. System (dashed curves) with cycling endpoint (solid curves): (*a*) End point moves slowly so that system closely tracks the end point; (*b*) end point moves too fast for system to make any response; (*c*) system can respond fast enough to chase the moving end point but not fast enough to closely track it.

6.2 OPEN SYSTEMS AND STEADY STATE

compared with the rate at which the end point is changing, the system catches up and "tracks" the end point closely, as in Figure 6.8a. If the response is very slow compared with the changing of the end point, the system gets to some average end point and pretty much stays there. That is, before the system has responded to one change, it is asked to change its direction of response. The result is that the system just stays comfortably in the middle (Figure 6.8b).

Between these two extremes, there is the case in which the speed of the system response and that of the end point change are of about the same order of magnitude. In this case, the system makes a "good show" of chasing the end point (Figure 6.8c). It never quite catches up but follows the end point cycling with the same frequency (cycles per unit time), with a certain amount of lag and possibly with decreased amplitude.

6.2. OPEN SYSTEMS AND STEADY STATE

In Figure 6.9, the system of the last section is "opened" to allow exchange with the environment. In order to keep the mathematical formulation simple enough to allow the conceptual principles to be visible, we assume that environmental region A is a source region within which concentration is constant, and that transfer through the barriers at a and c can only take place from left to right.

Now, we apply the same rules developed in subsection 5.5b and let

$k_{i,j}$ = rate parameter for transfer from i to j

The resulting equations are

$$\frac{dX_1}{dt} = k_{A,1}C_A - k_{1,2}C_1 + k_{2,1}C_2 \qquad (6.15a)$$

$$\frac{dX_2}{dt} = k_{1,2}C_1 - (k_{2,1} + k_{2,B})C_2 \qquad (6.15b)$$

A system is said to be in a *steady state* if all of the characteristics of the system (i.e., the state variables) remain constant with respect to time. To put it another way, the point that represents the system in state space is stationary with respect to time; it stays in one place. Such a state is also called a *stationary state*. In the example of Section 6.1, the steady state was the equilibrium state with $C_1 = C_2$. In order to determine whether such a steady state exists for the system of Figure 6.9, we can set the derivatives in

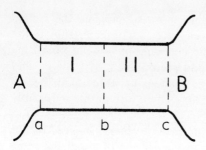

Figure 6.9. A two-compartment system open to environmental regions A and B.

equations (6.15) to zero and solve for C_1 and C_2. That is, for given parameters and environmental conditions (determined by the value C_A), we look for values of C_1 and C_2 that simultaneously satisfy

$$k_{A,1}C_A - k_{1,2}C_1 + k_{2,1}C_2 = 0 \qquad (6.16a)$$

$$k_{1,2}C_1 - (k_{2,1} + k_{2,B})C_2 = 0 \qquad (6.16b)$$

The solution to equations (6.16) gives the steady-state values,

$$C_1^{ss} = \frac{k_{A,1}(k_{2,B} + k_{2,1})}{k_{1,2}k_{2,B}} C_A \qquad (6.17a)$$

$$C_2^{ss} = \frac{k_{A,1}}{k_{2,B}} C_A \qquad (6.17b)$$

The characteristics of steady-state systems are more fully discussed in Chapter 7. Here, we want to bring out the following points:

i. The position of the steady state (i.e., the steady-state values of the state variables) is determined by the values of the system parameters together with the relevant characteristics of the environment.
ii. If C_1 and C_2 both have the values given by expressions (6.17)—that is, if $(C_1, C_2) = (C_1^{ss}, C_2^{ss})$—then they remain constant unless perturbed by some force not explicitly accounted for by the model.
iii. If $(C_1, C_2) \neq (C_1^{ss}, C_2^{ss})$, then the direction the system takes in state space is toward the steady-state point (C_1^{ss}, C_2^{ss}). This can be shown by direct substitution into the differential equations (6.16). That is, if $C_1 > C_1^{ss}$, then it is found that X_1 is decreasing; if $C_1 < C_1^{ss}$, then X_1 is increasing. An analogous statement applies to X_2.
iv. If the system is at steady state, and the value of C_A is changed, the

position of the steady state in the state space shifts and the system strives toward the new steady-state position. The discussion about chasing a moving end point in subsection 6.1d applies.

6.3. FLOW IN A GENERAL COMPARTMENTAL FRAMEWORK

The word "general" as used in the title of this section normally connotes *potential* applicability to a broad class of situations and specific applicability to none. The *type* of model that is discussed in this section has been applied in a very wide variety of situations, simple and complex, small and large. Examples were discussed in Section 2.4.

A compartmental framework may be useful for studying the flow or transfer of some entity through a system. In such a framework, each individual unit or increment of the entity is assumed to be fully described by specifying the compartment it is in. The state of the overall system is described by specifying the number of individuals within each such compartment. Changes of the system state are accounted for by the book-balancing equation (2.19), which is repeated here for reference:

(change in compartment j) = (sum of all transfers into compartment j)

$\qquad$ − (sum of all transfers out of compartment j)

$\qquad$ + (creation within compartment j)

$\qquad$ − (destruction within compartment j) $\qquad$ (6.18)

In this equation, the nature of the creation and destruction terms depends upon the entity whose flow is being studied. If it is total mass or total energy, then (except for nuclear processes) they are both zero by the laws of conservation of mass and energy. If the entity is matter of a certain *type*, they would involve chemical transformations; if energy of a certain type, they would involve thermodynamic interconversions between forms of energy; if living organisms, then they would involve the processes of birth and death.

Although the term "compartment" suggests an actual physical region of space, it is possible to think of more abstract types of compartments, as was pointed out in Section 2.4. Not every system can be prescribed in terms of a finite number of homogeneous compartments.

The questions that one is normally concerned with when such a description is used, relate to the rate of change of amounts within compartments and rates of transfer between compartments. The details depend on the following questions: i) which compartments trade with each other; ii) which compartments trade with the environment (anything that's not the system is automatically the environment of the system); and iii) how the trading takes place. Figure 2.8 shows, for example, a three compartment system with all trade routes indicated. In the three compartment system, each compartment can give material to two others and receive material from two others. Now, let

$r_{i,j}(t)$ = the absolute rate (amount per unit time) at which substance moves from compartment i to compartment j at time t

$Q_i(t)$ = net rate at which substance is produced within compartment i (it is actually production minus destruction) at time t

$X_i(t)$ = the amount of substance in compartment i at time t

$r_{i,e}(t)$ = rate at which substance is lost to the environment from compartment i at time t

$r_{e,i}(t)$ = rate at which the substance enters compartment i from the environment at time t

Then by the simple bookkeeping equation (6.18), the equation for the ith compartment would be (assuming that conditions are appropriate for the use of derivatives)

$$\frac{dX_i}{dt} = \sum_{\substack{j=1 \\ j \neq i}}^{N} r_{j,i}(t) - \sum_{\substack{j=1 \\ j \neq i}}^{N} r_{i,j}(t) + r_{e,i}(t) - r_{i,e}(t) + Q_i(t) \qquad (6.19)$$

In this equation, the first term on the right-hand side is the total rate of receipts from all other compartments. The second term is total losses to other compartments. The next two terms represent exchange with environment. The last term is net creation within the compartment and may be positive or negative (or zero, of course). Since there are N compartments, equation (6.19) is really a set ($i = 1, 2, \ldots, N$) of N simultaneous equations that together describe what's going on in the whole system.

Equation (6.19) is general in that it allows for the inclusion of $2N^2$ transfer pathways ($2N$ for each compartment) and N net creation processes (one for each compartment) and makes no commitment as to the nature of these processes. Delineation of the pathways and processes that are operative is usually the problem that must be addressed immediately after defining the compartments. An often-used simplification is to assume

6.3 FLOW IN A GENERAL COMPARTMENTAL FRAMEWORK

that the transfer pathways obey linear rate laws. That is,

$$r_{i,j}(t) = k_{i,j} X_i(t) \quad (6.20a)$$

$$r_{i,e}(t) = k_{i,e} X_i(t) \quad (6.20b)$$

$$i = 1, \ldots, N$$

$$j = 1, \ldots, N$$

As usual, the use of linear rate laws carries with it the assumption that the individual units each have equal chance of participating in the process (exiting from the ith compartment along a given pathway) and that with respect to this process they behave independently of each other; that is, that the individual units are stochastically identical and independent.[1] Usually the parameters $k_{i,j}$ and $k_{i,e}$ are taken to be constant, but in more elaborate treatments, they may be considered to depend upon other characteristics of the system that change as time goes on—as the permeability of a cell membrane may change with the age of the cell or as the ability to cross a river may depend upon the flood stage.

In Section 6.1, the parameters $k_{i,j}$ were taken to be inversely proportional to the volume of compartment i (that is, $k_1 = k_{1,2} = k/V_1$ and $k_2 = k_{2,1} = k/V_2$) with the same proportionality constant used for transfer in either direction. This followed from auxiliary assumption i of that section. In the more general situations now being considered, no such assumption is made. Indeed, it often happens that barriers are easier to cross in one direction than in the reverse. Prevailing winds, for example, may assist a bird in crossing a mountain range in one direction, and inhibit the return.

Using relations (6.20) with equation (6.19) gives

$$\frac{dX_i}{dt} = \sum_{\substack{j=1 \\ j \neq 1}}^{N} k_{j,i} X_j(t) - \sum_{\substack{j=1 \\ j \neq 1}}^{N} k_{i,j} X_i(t) - k_{i,e} X_i(t) + r_{e,i}(t) + Q_i(t) \quad (6.21)$$

In this formulation, the state of the system at any given time might be specified by the amount in each compartment. Thus, $X_1, X_2, \ldots, X_N$ are the state variables. The parameters for the system are the $k_{i,j}$ (there are

---

[1] Unfortunately, it is not possible to make a statement about a generally useful form for the terms $r_{e,i}$ and Q_i. Each depends too specifically upon the nature of the environment and its interaction with the system (for $r_{e,i}$) and upon the characteristics of the system itself (for Q_i).

$N(N-1)$ of them), the $r_{e,i}$ (there are N of them) and the Q_i (there are N of them).

Equation (6.21) may be put in terms of concentrations by dividing and multiplying each term by the appropriate volume (or area if that is more appropriate). That is, if we use a new set of parameters,

$$k'_{i,j} = k_{i,j} V_i$$

we can write

$$k_{i,j} X_i = k_{i,j} V_i \frac{X_i}{V_i}$$

$$= k'_{i,j} C_i$$

Doing this for each appropriate term in equation (6.21) gives

$$V_i \frac{dC_i}{dt} = \sum_{\substack{j=1 \\ j \neq i}}^{N} k'_{j,i} C_j(t) - \sum_{\substack{j=1 \\ j \neq i}}^{N} k'_{i,j} C_i(t) - k'_{i,e} C_i(t) + r_{e,i}(t) + Q_i(t) \quad (6.22)$$

In this formulation, the state variables are $C_1, C_2, \ldots, C_N$. The parameters, besides the $r_{e,i}$ and Q_i are the $k'_{i,j}$ [$N(N-1)$ of them] and the V_i (N of them).

In addition to the description of each compartment as given by equations (6.21) and (6.22), we can get an overall balance equation for the whole system. Using the general notation of equation (6.19), this would be

$$\frac{dX_T}{dt} = \sum_{i=1}^{N} r_{e,i} - \sum_{i=1}^{N} r_{i,e} + \sum_{i=1}^{N} Q_i \quad (6.23)$$

The description of compartmental systems developed in this section has remained fairly general. What has been developed is a framework for the construction of specific models for systems of a certain type. In actual practice, the description is made more specific as well as simpler by using the knowledge available about the system under study. We may know (or be willing to assume) that only certain compartments exchange with the environment and that only certain compartments exchange with each other. As a result, only relatively few of the terms are nonzero and have to be considered. The key steps in the modeling process might be summarized

6.3 FLOW IN A GENERAL COMPARTMENTAL FRAMEWORK

as follows:

i. Decision as to what constitute the compartments;
ii. Decision as to which of all the possible transfer routes are actually used;
iii. Determination of the appropriate rate laws for these transfer routes. This step often is accomplished using the linear rate law assumptions of (6.20).

If the closed two-compartment system considered in Section 6.1 is described in terms of equation (6.22), we get

$$\frac{dX_1}{dt} = V_1 \frac{dC_1}{dt} = k'_{2,1} C_2 - k'_{1,2} C_1$$

Note that the expression on the right-hand side cannot be factored to give an equation of the form of (6.3) unless $k'_{2,1} = k'_{1,2}$.

EXERCISE 6.1. Cite, from your own field of biology, a system that can be described using a compartmental model. Formulate (and justify) a description of the individual compartments. Formulate (and justify) the designation of the allowable transfer pathways. Which of these allowable transfers may be assumed to follow linear rate laws?

EXERCISE 6.2. Cite, from your own field of biology, a system for which a compartmental description would not be acceptable. Why not?

An important tool in the experimental study of compartmental systems involves the introduction of known amounts of labeled material into the system. The material introduced is identical to that which is naturally present, in that the system cannot distinguish between the two, but it is labeled so that the investigator can distinguish between them. The observations in such an experiment consist of measurements of amounts and concentrations of labeled material in the compartments, and of the ratio of labeled material to naturally occurring material. Application of equations of the type of (6.21) and (6.22) may then allow estimates of the parameters insofar as they apply to the labeled material. The same parameter values are assumed to hold for the natural operation of the system.

The conjunction of label techniques with compartmental decompositions is widely applied in biological science. When the material being studied is molecular, the label often consists of radioactive or other isotopic tracers (Atkins, 1969; Jacquez, 1972; Shipley and Clark, 1972); in ecological studies, the labels consist of bands, tags, or other markings applied to animals and are referred to as the *mark-recapture* technique (Southwood, 1966). It is important to bear in mind that the interpretation of the results of such studies is very sensitive to the assumptions made about the system and about the relation between the labeled and the natural material.

6.4. DIFFUSION AND MIGRATION IN A "CONTINUOUS" SYSTEM

The compartmental type of model is based on the assumption that the system can be divided into compartments, each of which has a certain degree of homogeneity. Although changes with respect to time may sometimes be treated as taking place continuously, any changes with respect to position can occur only in going from one compartment to another. If the compartments are contiguous physical regions, then we may say that the changes take place across the boundaries between regions.

It is clearly not always useful to divide the system into distinct regions. In this section, we switch the focus to changes that may be considered to be continuous with respect to position in the system. This type of model and the resulting formalism has been applied in many areas of biology. A few examples are random spread of biological populations, disease spread and epidemiology, spread of water and air pollution, and intracellular metabolite movement. After developing the basic description, we reexamine some of the underlying assumptions.

6.4a. The Rate of Flow

To begin simply, we assume that any gradient (and, therefore, any net diffusion) is in only one direction, say the X direction. The first question we ask is this: How fast is the entity of interest moving past some particular point, say point x in Figure 6.10? To answer this, we need to construct a model of what's going on.

The first part of the model is a specification of the underlying behavior of the individual particles. As in the case of transfer across barriers, we assume that the particles are stochastically independent and identical and the space is inherently homogeneous. In particular, a given particle is as likely to be one place as any other and is as likely to be moving in one direction as in any other.

The next part of the model is really designed to help reformulate the question in a more precise form. We imagine a two-dimensional rectangle

6.4 DIFFUSION AND MIGRATION IN A "CONTINUOUS" SYSTEM   165

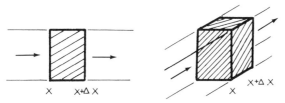

Figure 6.10. Model for derivation of the diffusion equations.

at point x, oriented to be perpendicular to the direction of net flow. For convenience, we construct the rectangle to have one unit of area. Now, if in some time interval Δt, there is more flow to the right than to the left, then the net flow must be said to be toward the right. To put it another way, we can take the rectangle that has been constructed and take it to be an imaginary barrier. The direction and magnitude of the net flow at point x is determined by the different rates of flow across the imaginary barrier in the positive and negative directions. The next step is to examine this flow in detail.

Being a two-dimensional figure, this rectangle has zero thickness. However, let's imagine that it has finite thickness Δx, as shown in Figure 6.10. This gives a "plane" of unit cross section with thickness Δx, so the volume is *numerically* equal to Δx. We are going to do a little figuring with this "thick plane," and then let the thickness shrink back to zero.

Next, we need a clear definition of "net flow" as a measure of the excess of the amount traveling in one direction over the amount traveling in the opposite direction. In order to get such a measure, we use the old bookkeeping equation (6.18) applied to the region between x and $x+\Delta x$.

We suppose that Δx is so small that no particle has time to change direction within the region x to $x+\Delta x$. This is not *too* drastic, since Δx is eventually going to be shrunk to zero anyway,[2] and it allows the further development to be very much simplified. The result of this assumption is that any particle traveling to the right *within* the region had to enter from the left, and any particle traveling to the left within the region had to enter from the right. Recall that for any given particle, one direction is as likely as another. This means that the number of particles entering from the left is proportional to the concentration at point x. Similarly, the number of particles entering from the right is proportional (same proportionality constant) to the concentration at $x+\Delta x$. If $C(x) > C(x+\Delta x)$, then the net

[2]However, see the discussion in subsection 6.4c.

flow is toward the right, and its mangitude is

$$\text{net flow to right through } \Delta x = D[C(x) - C(x+\Delta x)] \quad (6.24)$$

where D is a proportionality constant that depends on the inherent mobility (speed of movement) of the particles, and anything that may influence this mobility. In molecular systems, it would be called the *diffusion coefficient*.

Expression (6.24) gives the total amount (under our idealized conditions) flowing through the volume Δx. Obviously, if we shrink Δx down to zero, this total amount is going to shrink to zero. What is needed is the *average* flow—something that does not depend on the particular choice of Δx. The average flow (per unit of volume) would be $D[C(x) - C(x+\Delta x)]/\Delta x$. Equation (6.24) becomes

$$\frac{\text{net flow to right through } \Delta x}{\Delta x} = D \frac{C(x) - C(x+\Delta x)}{\Delta x}$$

The amount flowing per unit of volume *at the point* x would be obtained by taking the limit as $\Delta x \to 0$. If you look at the fraction on the right-hand side, you see that the limit is just the negative of the derivative of C. So, if we let $J(x)$ stand for the flow per unit time flowing across a rectangle of unit area (that is, it is amount per unit time per unit of area) we get

$$J(x) = -D\frac{dC}{dx} \quad (6.25)$$

The negative sign is completely in accord with the idea that material should be flowing from high concentration to low concentration. If dC/dx is positive, it means that concentration is increasing as we go to the right and the net flow would be to the left—opposite to the concentration gradient. Similarly, a negative value of dC/dx means concentration decreasing to the right and, therefore, a positive flow to the right. If $dC/dx = 0$, then the concentration is uniform and the system is at equilibrium—the flows to the left and right are in balance and there is no net flow.

6.4b. Rate of Change at a Given Point

A companion question to "How fast is something moving?" is the question, "How fast is its density changing at some particular spot?" This

6.4 DIFFUSION AND MIGRATION IN A "CONTINUOUS" SYSTEM 167

question can also be attacked using Figure 6.10. We go back to the region of unit cross section and nonzero length Δx, but now we know how to define the flow at the point x and also at the point $x + \Delta x$. The question being asked now is How fast is the concentration changing at the point x? We look first at how the concentration is changing in the region between x and $x + \Delta x$.

i. Using equation (6.18), we get

$$\text{change in amount in the region} = \text{flow in} - \text{flow out}$$
$$+ \text{net production}$$
$$= J(x) - J(x + \Delta x) + Q$$

where Q is total production in the region.

ii. The total amount in the region can also be expressed as the average concentration (call it $\overline{C}$) times the volume (*numerically* equal to Δx). Using the result of step i, we get for a fixed Δx

$$\Delta x \frac{d\overline{C}}{dt} = J(x) - J(x + \Delta x) + Q \tag{6.26}$$

iii. Of course, as $\Delta x \to 0$, the total amount in the region always goes to zero. We need, therefore, to convert to amount per unit volume

$$\frac{d\overline{C}}{dt} = \frac{J(x) - J(x + \Delta x)}{\Delta x} + \frac{Q}{\Delta x} \tag{6.27}$$

The last term on the right-hand side is a "production density," that is, rate of production per unit volume.

iv. Assuming that we can take the limit as $\Delta x \to 0$ (this is discussed later), we get the following results:

a. $\overline{C} \to C(x)$

b. $\dfrac{J(x) - J(x + \Delta x)}{\Delta x} \to -\dfrac{dJ(x)}{dx}$

c. $Q/\Delta x$ approaches what might be called the "production density" at x, which will be called $q(x)$.

The result is

$$\frac{dC(x)}{dt} = -\frac{dJ(x)}{dx} + q(x) \qquad (6.28)$$

v. At this point, we can use the result of equation (6.25) to get

$$\frac{dJ(x)}{dx} = \frac{d}{dx}\left(-D\frac{dC(x)}{dx}\right)$$

$$= -D\frac{d^2C(x)}{dx^2}$$

Putting this into (6.28), we wind up with

$$\frac{dC(x)}{dt} = D\frac{d^2C(x)}{dx^2} + q(x) \qquad (6.29)$$

Let's see what this equation tells us physically. For the moment, let's assume $q(x)\equiv 0$. We know, first of all, from equation (6.25) that if we have no concentration gradient there will be no net flow. It's also clear that there will be no changes with respect to time (except for the $q(x)$ term). If the no-flow-no-change condition pertains to *every* point x, then the system is completely at equilibrium with respect to the entity of interest.

Now suppose that $dC(x)/dx < 0$ so that there is a net flow to the right, and suppose that

$$\frac{d^2C(x)}{dx^2} = 0$$

so that dC/dx is constant and the flow remains the same from point to point. The amount flowing into any region is the same as the amount flowing out of the region, and, therefore, assuming $q(x)\equiv 0$, the amount in the region remains constant with respect to time; its time derivative is zero.

Now suppose that $dC/dx < 0$ and $d^2C/dx^2 < 0$. Both C and its "rate of change" have a negative "rate of change" with respect to x. The flow is toward the right and becomes faster as we go toward the right. Therefore, given any region such as that shown in Figure 6.10, the flow out ($J(x+\Delta x)$) is greater than the flow in ($J(x)$), and the total amount in the region is decreasing ($dC/dt < 0$).

6.4 DIFFUSION AND MIGRATION IN A "CONTINUOUS" SYSTEM 169

EXERCISE 6.3. Make a physical analysis of equations (6.25) and (6.29) (similar to that in the preceding paragraphs) for all combinations of dC/dx and d^2C/dx^2 positive, negative, and zero. Assume $q(x) \equiv 0$. Make liberal use of drawings and graphical representations.

EXERCISE 6.4. Do a dimensional analysis of equations (6.25) and (6.29). In particular, what are the dimensions of J and D?

Note that in analyzing the physical content of a differential equation, it is generally useful to consider how the equation was "born." A differential equation always comes to life through some process of taking a limit. If we back off from the limit and magnify our "ratio of infinitesimally small differences" so that dC/dx is read as $\Delta C/\Delta x$, it often helps in the understanding of the "spirit" of the equation.

6.4c. The Assumptions

As in the discussion of diffusive transfer between compartments, it is assumed that the individuals or particles are identical in a stochastic sense and behave independently of each other. As before, the independence assumption may break down if the system is concentrated or dense enough so that the individuals begin interacting to the point of influencing each other's probability of migration. In such a case, the degree of interaction depends on the density (or concentration), and the simple linear rate laws no longer hold. The nature of the departure from linearity depends, of course, on the nature of the interactions.

Also, as before, it is assumed that numbers of things and of events can be used instead of the expectations of these numbers. Recall that this generally finds justification on the basis of a coefficient of variation that decreases as the expectation rises, and the assumption that the expectation is high enough so that the coefficient of variation is very low.

There are the same problems as before in the taking of derivatives with respect to time. It must be assumed that things are changing slowly enough so that the difference ratio becomes essentially constant while Δt is still large enough to accommodate a "large" number of events. As before, we must not ask the equations for information that requires direct consideration of details on too fine a time scale.

An additional problem arises in the taking of derivatives with respect to position. The same considerations hold as for time derivatives. Changes with respect to position have to be taking place slowly enough so that the ratios (including the ratio $Q/\Delta x$) become essentially constant while Δx is yet large enough to accommodate a "large" number of individuals. To put it another way, the concentration gradient must not be so steep that appreciable changes in concentration occur over distance comparable to the average distance between individuals. Furthermore, we must be careful about not asking the equations for information that requires detail on such a fine spatial scale.

6.4d. Flows in More Than One Direction

Since the discussion in this subsection involves motion in more than one direction, the symbols X, Y, and Z are used to designate directions; the symbols x, y, and z are used for the coordinate values in the respective directions. In most real life situations, the traveling individuals don't line their directions of travel up with our arbitrary choice of a coordinate system. However, if an individual or particle is traveling in any arbitrary direction, its motion can be *described* by saying that it is moving at a certain rate in the X direction, a certain rate in the Y direction, and a certain rate in the Z direction.[3] Just as the position of the particle at any time t can be described by an ordered triple (x,y,z), so the velocity of the particle can be described by an ordered triple $(\partial x/\partial t, \partial y/\partial t, \partial z/\partial t)$, consisting of the velocities in the three coordinate directions. Similarly, the net flow would need to be expressed as an ordered triple (J_x, J_y, J_z) giving the flow in the three coordinate directions.

This leads to the conclusion that equation (6.25) is really an equation for J_x. A similar equation could be derived in the same manner for flow in the Y and Z directions. The resulting three equations can be summarized in vector notation, letting $\mathbf{J} = (J_x, J_y, J_z)$,

$$\mathbf{J} = (J_x, J_y, J_z) = -\left(D_x \frac{\partial C}{\partial x}, D_y \frac{\partial C}{\partial y}, D_z \frac{\partial C}{\partial z}\right) \qquad (6.30)$$

If attention is focused on the change in concentration at some point p, it must be realized that the change occurs because of flows in three directions. The result is

$$\frac{dC(p)}{dt} = D_x \frac{\partial^2 C}{\partial x^2} + D_y \frac{\partial^2 C}{\partial y^2} + D_z \frac{\partial^2 C}{\partial z^2} + q(p) \qquad (6.31)$$

---

[3] The coordinate axes are conceptual constructs that we have created so that we can more easily describe some aspect of reality. The moving individual does not care where we have drawn the coordinate axes or how we describe its motion.

6.5 CHEMICAL INTERACTIONS AND EQUILIBRIUM

If the mobilities (diffusion coefficients) in the three directions are the same, equations (6.30) and (6.31) could be written

$$\mathbf{J} = -D\left(\frac{\partial C}{\partial x}, \frac{\partial C}{\partial y}, \frac{\partial C}{\partial z}\right)$$

$$\frac{dC(p)}{dt} = D\left(\frac{\partial^2 C}{\partial x^2} + \frac{\partial^2 C}{\partial x^2} + \frac{\partial^2 C}{\partial z^2}\right)$$

EXERCISE 6.5. Construct an example in your own field that meets the assumptions of this section so that flow or migration can be described by equation (6.30) and changes in concentration or density can be described by equation (6.31). In constructing the example,
a. Define the quantity being considered.
b. Define what is meant by its concentration or density.
c. What is the production term $q(p)$?
d. Define the spatial scale and time scale on which the use of the derivatives is justifiable.
e. Can the coefficients D_x, D_y, and D_z be considered to be equal?

6.5 CHEMICAL INTERACTIONS AND EQUILIBRIUM

In Section 5.7, chemical systems are used to develop a formulation for the interaction between individuals in a population. In such systems, the products of reaction may themselves undergo reaction to reform the reactants. The *net* rate of reaction is then the difference between the rates of the forward and reverse reactions. Equation (5.41) must, therefore, be rewritten

$$\frac{1}{a}\frac{dC_A}{dt} = \cdots = -kC_A^a C_B^b C_C^c \cdots + \text{rate of reverse reaction} \quad (6.32)$$

Under the assumption that all molecules behave independently, except for explicitly considered interactions, the reverse reaction proceeds independently of the forward reaction and follows the same rules. If we modify the general reaction of subsection 5.7c to include the products and the reverse reaction, we have

$$aA + bB + cC + \cdots \rightleftharpoons mM + pP + \cdots \quad (6.33)$$

Using k_f and k_r to denote the rate constants for the forward and reverse reactions,

$$\text{net rate of reaction} = k_f C_A^a C_B^b C_C^c \ldots - k_r C_M^m C_P^p \ldots \qquad (6.34)$$

The net rate may be either positive or negative. A positive net rate indicates that the reaction is proceeding to the right *as it is written*. A negative net rate indicates that it is proceeding to the left *as it is written*.[4]

The rate of change of the concentrations of the individual species is (using the same reasoning as in Section 5.7)

$$\frac{1}{a}\frac{dC_A}{dt} = \frac{1}{b}\frac{dC_B}{dt} = \cdots = -\frac{1}{m}\frac{dC_M}{dt} = -\frac{1}{n}\frac{dC_P}{dt} = \cdots$$

$$= -k_f C_A^a C_B^b C_C^c \ldots + k_r C_M^m C_P^p \ldots \qquad (6.35)$$

[*Note*: Be sure you understand the arrangement of signs in (6.35).]

Now, if reaction (6.33) is taking place in a closed system (nothing added nor removed), the concentration of reactants is reduced, whereas that of products is increased (if net reaction is to the right). Therefore, the rate of the forward reaction is reduced, whereas that of the reverse reaction is increased. As time goes on, the two rates approach equality, at which point the net rate is zero, and the system is said to be at equilibrium,

$$k_f C_A^{*a} C_B^{*b} \ldots = k_r C_M^{*m} C_P^{*p} \ldots \qquad (6.36)$$

The asterisks here are meant to show that (6.36) does not hold for just any values of the concentrations, but only for the equilibrium values. Dividing both sides of (6.36) by $C_A^{*a} C_B^{*b} \ldots$ and also by k_r, we get

$$\frac{k_f}{k_r} = \frac{C_M^{*m} C_P^{*p} \ldots}{C_A^{*a} C_B^{*b} \ldots} \qquad (6.37)$$

Under the assumptions that have been made, the left-hand side of this equation is a ratio of constants and must, therefore, be a constant. By the word "*constant*," we mean that it does not depend on the concentrations of any of the substances involved. It may, of course, depend on other things such as temperature and pressure. However, neither the numerator nor the

---

[4]The as it is written specification is important since (6.33) may also be written

$$mM + pP + \cdots \rightleftarrows aA + bB + cC \cdots \qquad (6.33')$$

6.6 DIRECT COMPETITION EFFECTS IN POPULATION MODELS   173

denominator on the *right-hand* side can be considered "constant." Certainly the equilibrium concentrations depend on how much material we started with. That is, these concentrations must depend on the total concentration of material in the system. What equation (6.37) says is that at equilibrium, the ratio of the expression for product concentration to that of reactant concentration does not depend on the total amount of material in the system. It is a "constant," and is called the equilibrium constant for the reaction under consideration.[5]

6.6. DIRECT COMPETITION EFFECTS IN POPULATION MODELS

In this section, we amend the simple model of population change developed in Section 5.2, by adding a term for direct competition between individual members of the population. As before, the word *"population"* is being used in a general sense, although the most direct applications of the formalism are with respect to populations of organisms.

It seems reasonable to say that if the population density is so low that two individuals never meet, then there is no direct competition effect; a "competition event" occurs only when individuals meet. In other words, the competition effect is, to a beginning approximation, taken to be proportional to the probability that two individuals encounter each other. Under the assumption that movements of individuals are stochastically independent and identical, this probability is proportional to N^2 (Section 5.7). Since the competition effect is taken to be adverse, equation (5.9) is amended to

$$\frac{dN(t)}{dt} = k_1 N(t) - k_2 N(t)^2 \qquad (6.38)$$

where k_1 and k_2 are both taken to be positive.

To determine how such a population behaves, we do a little qualitative reasoning with this equation. When N is small, the competition effect is small. That is, the relative values of k_1 and k_2 are such that for small N, $k_1 N \gg k_2 N^2$. Therefore, for small N, equation (6.38) behaves approximately as equation (5.9).

If N is increased, however, it must eventually get large enough so that

---

[5]CAUTION. If the reaction is written as in (6.33'), then the equilibrium constant is the reciprocal of that for (6.33). Convenience alone dictates the choice of direction, but once a choice is made, it must be used consistently.

$k_1 N \ll k_2 N^2$. Equation (6.38) then becomes

$$\frac{dN}{dt} \cong -k_2 N^2, \qquad \text{large } N \qquad (6.39)$$

If N were to be this large, competition would force a decline in the population. The rate of decline would be proportional to the square of the population size. Integrating equation (6.39), we get

$$\frac{1}{N(t)} - \frac{1}{N(t_0)} = k_2(t - t_0) \qquad (6.40)$$

Since $1/N_0$ is a constant, the result is that as time goes on $(t - t_0$ gets large), $1/N(t)$ gets large, and, therefore, $N(t)$ gets small. The relation between N and $k_2 t$ is a reciprocal one and has a shape qualitatively similar to that in Figure 6.11.

Figure 6.12 shows the dependence of the derivative dN/dt on N. For very small N, it would be a straight line (line a). For very large N, it would be negative and look like solid line b. In between, where neither term is dominant, the two parts would be connected (in this case) by a smooth curve as shown by the dashed line. This line crosses the axis for $dN/dt = 0$, for some value of N. This value of N is a steady-state value, which is denoted N^{ss}. For this value of N, we have

$$k_1 N^{ss} = k_2 N^{ss^2}$$

so that

$$N^{ss} = \frac{k_1}{k_2} \qquad (6.41)$$

At this value of N, the competition effects and the growth effects just

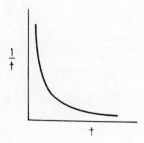

Figure 6.11. A variable and its reciprocal.

6.6 DIRECT COMPETITION EFFECTS IN POPULATION MODELS

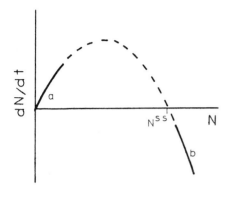

Figure 6.12. Behavior of dN/dt as a function of N from equation (6.38).

balance each other. For $N > k_1/k_2$, the population is predicted to decrease, whereas for $N < k_1/k_2$, the population is predicted to increase. For $N = k_1/k_2$, the population is predicted to remain constant.

If we start with a low but nonzero value for N, the foregoing arguments suggest that the system initially increases exponentially so that its time curve looks like solid curve a in Figure 6.13. Eventually, N will become constant as it approaches k_1/k_2 (solid line b). Connecting these two types of limiting behavior by a smooth curve, we get a picture of the qualitative behavior of N as a function of time—without actually having solved the differential equation (6.38). This type of curve is often referred to as a "logistic curve."

Notice that the last part of the curve resembles that of Figure 6.3, in that it is asymptotically approaching a limiting value. It is reasonable to say that the limit is approached as the carrying capacity of the environment (represented by k_1/k_2) is saturated. Of course, changes in the environment

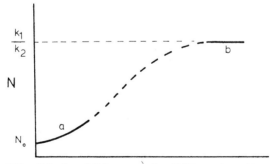

Figure 6.13. Behavior of N from equation (6.38), starting with small N_0.

may alter k_1/k_2, so that the asymptote is changed. The system would then attempt to adjust to the new end point. If the end point is continually changing, then the system has the "problem" of chasing a moving end point. The principles discussed in subsection 6.1d apply.

The differential equation (6.38) is often written in terms of departure from the carrying capacity; using expression (6.41), it may be rearranged to the form

$$\frac{dN}{dt} = k_1 N \left(\frac{N^{ss} - N}{N^{ss}} \right)$$

In this form, the intrinsic rate of increase, given by the factor $k_1 N$, is moderated by the degree to which the carrying capacity has been saturated. By developing the equation in the form of (6.38), we have been able to show how this moderation arises from the interactions within the population. In particular, we are able to clearly see that the equation does *not* account for a variety of effects such as age structure, cohort formation, and territoriality. These effects are briefly discussed in Appendix A5.

6.7. MORE GENERAL INTERACTION MODELS

Section 6.6 discusses competition in terms of the probability of encounters between individuals. However, such encounters could lead to cooperative rather than competitive effects. For example, one bird may be alerted by the danger signal given by another. If the kind of assumptions are made as in Section 5.7, especially regarding independence of the individuals, then an equation incorporating cooperative effects would be analogous to equation (6.38),

$$\frac{dN(t)}{dt} = k_1 N(t) + k_2 N(t)^2 \tag{6.42}$$

with $k_2 > 0$.

If $k_1 < 0$ in (6.42), then we have the circumstance that maintenance of the population *depends* on the cooperation. In such a case, if the population were ever to be so small that $|k_1 N| > |k_2 N^2|$, then the rate of change would become negative, and it would die off. That is, such a species would require a minimum population size for survival.

Note that (6.42) does not have any provision for placing an upward limit

6.7 MORE GENERAL INTERACTION MODELS

on the population size when $k_2 > 0$. One could reason, however, that as the population continues to increase, competition effects would begin to outweigh cooperative effects. A competition term proportional to N^3 might, therefore, be added. For large enough N, this term would dominate over the other two. Under the independence assumption, this term would be proportional to the probability that three individuals meet at the same time. It says, in effect, "two's company, three's a crowd."

The same kinds of concepts carry through to interaction between different species, say A and B. The system might be represented by two equations,

$$\frac{dN_A}{dt} = \kappa_A N_A + \kappa_{AB} N_A N_B \tag{6.43a}$$

$$\frac{dN_B}{dt} = \kappa_B N_B + \kappa_{BA} N_A N_B \tag{6.43b}$$

(Note the order of subscripts; κ_{AB} indicates the effect on A of B.) In these equations, the parameters may be positive or negative. They represent a variety of interaction effects, depending upon the signs.

a. Synergism: $\kappa_{AB} > 0$
                   $\kappa_{BA} > 0$
b. Competition: $\kappa_{AB} < 0$
                   $\kappa_{BA} < 0$
c. Prey–Predation (A = prey, B = predator):
                   $\kappa_{AB} < 0$
                   $\kappa_{BA} > 0$

The properties of a system described by equations (6.43) are further examined in the next chapter. We see that cases b and c provide a built-in limit on population size, but case a does not. Competition terms such as $\kappa_{AA} N_A^2$ and $\kappa_{BB} N_B^2$ could be added to provide such limits. The general signal–flow graph showing the relation between N_A, N_B, and their time derivatives is shown in Figure 6.14.

Equations (6.43) could be extended to include direct pairwise interactions involving any number of species. Numbering the species, $1, 2, \ldots, r$,

$$\frac{dN_i}{dt} = \kappa_i N_i + \sum_{j=1}^{r} \kappa_{i,j} N_i N_j \tag{6.44}$$

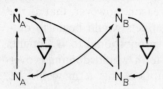

Figure 6.14. Signal-flow graph for two-species interaction. The triangles indicate time delays associated with integration (see Section 5.9).

6.8. INDIRECT INTERACTION; SATURATION OF A LIMITED RESOURCE

Suppose we have a process in which each participating individual requires the temporary use of a limited resource. In such a case, the rate of the total process may be limited by the saturation of the capacity of the limited resource. The individuals are not competing through direct interaction as in Sections 6.6 and 6.7 but are competing for the limited resource. An example might be the passage of grains of sand out of a narrow-necked bottle, the limited resource in this case being the limited space in the neck.

A classical biological example of such a "bottleneck" saturation effect is the Michaelis–Menten scheme for enzyme action. Letting

E = enzyme X = enzyme–substrate complex
S = substrate P = product

the mechanism is pictured as follows:

$$E + S \underset{-1}{\overset{1}{\rightleftarrows}} X \qquad (6.45a)$$

$$X \overset{2}{\to} P + E \qquad (6.45b)$$

where the reactions are numbered 1, −1, and 2.

The enzyme is regenerated in step 2 and so plays the role of a temporarily used resource in conversion of S to P. If conditions are such that $C_S \gg C_E$ (the usual laboratory condition), then E is also a "limited resource."

The object now is to find out how the rate of production of product ("velocity" of the reaction) depends on C_S and C_E. Usual notation is to let v be the reaction velocity,

$$v = \frac{dC_P}{dt}$$

6.8 SATURATION OF A LIMITED RESOURCE

Applying the law of mass action to (6.45),

$$v = k_2 C_X \tag{6.46}$$

$$\frac{dC_X}{dt} = \text{rate of formation} - \text{rate of breakdown}$$

$$= k_1 C_E C_S - (k_{-1} + k_2) C_X \tag{6.47}$$

At this point, we remember that the resource (enzyme) is limited and that C_E measures only the available resource. If C_E^0 is the total resource and C_X is the amount being used, we get

$$C_E^0 = C_E + C_X = \text{const.} \tag{6.48}$$

Using this relation to eliminate C_E in (6.47) gives

$$\frac{dC_X}{dt} = k_1 (C_E^0 - C_X) C_S - (k_{-1} + k_2) C_X$$

$$= k_1 C_E^0 C_S - (k_{-1} + k_2 + k_1 C_S) C_X \tag{6.49}$$

Equations (6.46) and (6.49) are a *system* of two equations that now describe our chemical system. Unfortunately, this system of equations is too hard to handle in all its generality. However, if both C_E^0 and C_S are kept constant, there is some value of C_X for which $dC_X/dt = 0$. If C_X is higher than this, then $dC_X/dt < 0$, and it decreases. If C_X is less, it increases. Thus if we wait long enough, then C_X gets to its *steady-state* value and stays there. In practice, for most enzymes as they are ordinarily studied in the laboratory, this point is reached very quickly.

We, therefore, assume that the steady state has been reached, and we ask how the characteristics of that steady state depend upon C_E^0 and upon C_S, that is, how the steady-state use of the resource depends upon its total amount and on the demand level. Setting equation (6.49) equal to zero and solving for C_X,

$$C_X = \frac{k_1 C_E^0 C_S}{k_{-1} + k_2 + k_1 C_s} \tag{6.50}$$

Substituting into (6.46) gives, finally, an equation that shows how the steady-state rate depends upon C_E^0 and C_S,

$$v = \frac{k_1 k_2 C_E^0 C_S}{k_{-1} + k_2 + k_1 C_S} \tag{6.51}$$

Now let's look at the properties of this equation. First, how does v depend on C_E^0? Equation (6.51) can be written

$$v = \frac{k_1 k_2 C_S}{k_{-1} + k_2 + k_1 C_S} C_E^0$$

Therefore, for a given value of C_S, the rate of the process is proportional to the amount of limited resource.

The dependence on C_S (for fixed C_E^0) is not so easy. To examine it, we look at the behavior for small C_S and for large C_S and then connect the two.

SMALL C_S. C_S can, in principle, be made so small that $k_{-1} + k_2 + k_1 C_S$ is not much different from $k_{-1} + k_2$, so equation (6.51) becomes

$$v = \frac{k_1 k_2 C_E^0}{k_{-1} + k_2} C_S, \quad \text{small } C_S \tag{6.52}$$

So, for small C_S, v is approximately proportional to C_S (line a, Figure 6.15). This is reasonable, since for small C_S, there is plenty of resource to go around. Each new increment of substrate has the same ample supply of available resource.

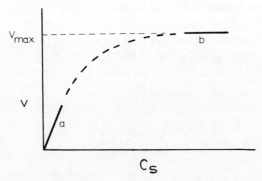

Figure 6.15. Saturation curve of v versus C_S from equation (6.51) or (6.54).

6.8 SATURATION OF A LIMITED RESOURCE

LARGE C_S. C_S could be made so large that

$$k_{-1} + k_2 + k_1 C_S \approx k_1 C_S$$

Equation (6.51) then becomes

$$v = k_2 C_E^0, \quad \text{large } C_S \tag{6.53}$$

Here, v doesn't depend on C_S at all. The resource is saturated (line b, Figure 6.15).

Connecting these limiting behaviors by a smooth curve, we get the curve showing how the *steady-state* rate depends on C_S for fixed C_E^0.

Note on Terminology. In the literature of enzyme kinetics, equation (6.51) is usually written in a slightly different form. From (6.53), we see that $k_2 C_E^0$ is the maximum rate that can possibly be attained and is written V or sometimes V_{max} to give

$$v = \frac{k_1 V_{max} C_S}{k_{-1} + k_2 + k_1 C_S}$$

Dividing top and bottom by k_1,

$$v = \frac{V_{max} C_S}{(k_{-1} + k_2)/k_1 + C_S}$$

The package of constants in the denominator is the "steady-state Michaelis constant,"

$$K = \frac{k_{-1} + k_2}{k_1}$$

so

$$v = \frac{V_{max} C_S}{K + C_S} \tag{6.54}$$

The usefulness of equation (6.54) as compared to (6.51) is that (6.54) is written in terms of the characteristics of the curve in Figure 6.15. The curve starts out with slope V/K and approaches the horizontal asymptote V_{max}. When $C_S = K$, then $v = (1/2) V_{max}$. Thus K may be interpreted without regard to the underlying phenomena simply as the concentration (or demand) needed for half-maximal rate.

Equations of the form of (6.54) are widely used as a basis for the description and interpretation of saturation phenomena. In Appendix A1, it is used in a description of enzyme action; in Appendix A2, it is used in a description of nutrient limited microorganism growth; it finds use in population ecology to describe resource-based competition (Titman, 1976). In equation (6.54), the demand (represented by substrate concentration) is allowed to vary, whereas the resource supply (represented by enzyme concentration) is held in limited supply. In some contexts, however, the thing that is in limited fixed supply is termed *demand*. In economic theory, the demand for certain commodities (food, for example) is limited but renewable. A plot of consumption versus resource supply, therefore, is a saturation curve, such as Figure 6.15. In this context, it is the economic demand that plays the role of "resource." In the discussion of Chapter 2 on dependence of food intake on availability of food, the saturation curve (Figure 2.9) can be interpreted in just such a way. For the purpose of applying equation (6.54), the question of which of two entities plays the role of supply and which of demand depends upon which of the two is limiting.

From equation (6.54), we can get the dependence of rate on substrate concentration,

$$\frac{dv}{dC_S} = \frac{V_{max}}{(C_S + K)^2} \qquad (6.55)$$

This equation has the general form of equation (5.46), which was used in subsection 5.8c as a description of decreasing sensitivity. That is, the rate of the process is sensitive to the amount of substrate, but this sensitivity diminishes as the total amount is increased.

6.9. COMBINING COOPERATION AND RESOURCE LIMITATION; THE HILL EQUATION

In this section, we consider the following general type of situation:

a. the process requires utilization of a resource whose total supply is constant;
b. utilization of the resource requires the cooperation of n individuals.

6.9 COOPERATION VERSUS RESOURCE LIMITATION

Such a description was at one time proposed as a model for the action of enzymes that seem to require simultaneous involvement of more than one substrate molecule. The description consists of the following two chemical reactions:

$$E + nS \underset{-1}{\overset{1}{\rightleftharpoons}} X \quad (6.56a)$$

$$X \overset{2}{\to} E + P \quad (6.56b)$$

In place of (6.46), (6.47), and (6.48), we would have

$$v = k_2 C_X$$

$$\frac{dC_X}{dt} = k_1 C_E C_S^n - (k_{-1} + k_2) C_X$$

$$C_E^0 = C_E + C_X = \text{const.}$$

The difference between this case and that considered in Section 6.8 is that here C_S is raised to the nth power. Carrying the development through as in Section 6.8 and using the same notation as for equation (6.54), we get

$$v = \frac{V_{\max} C_S^n}{K + C_S^n} \quad (6.57)$$

In this case, for small C_S, we have

$$v = \frac{V_{\max}}{K} C_S^n \quad (6.58a)$$

The cooperative effects dominate in this region and v shows an increasing sensitivity to C_S as its value increases (subsection 5.8b). A plot of v versus C_S in this region has the appearance of solid curve a in Figure 6.16. Depending upon the values of V/K and of n, it may even be that a minimum value of C_S is required before v becomes observable. In such a case, the observations may be described in terms of a lag region or of a required threshold value for C_S.

At the opposite extreme—for large C_S—the competition effect becomes dominant, and the behavior is just the same as in Section 6.8; the curve looks like b in Figure 6.16. A sigmoid curve is obtained by connecting the two regions.

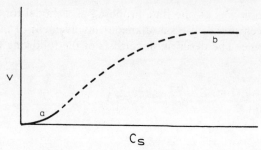

Figure 6.16. Curve according to equation (6.57), showing cooperative effect for small C_S and competitive effect for large C_S (compare with Figure 6.13).

In both this case, which involves indirect cooperation and competition, and in the case of Section 6.6, which involves direct interaction, a sigmoid curve results from the interplay of two opposing effects, each of which is dominant in different regions of the state space.

The qualitative behavior of Figure 6.16 is shown by many enzymes. Detailed investigations have resulted in explanatory models more sophisticated than the simple mechanism discussed in this section. Even so, this mechanism still serves well as a conceptual reference point, especially when new investigations are being launched.

6.10. SOME REMARKS ON CYCLING BEHAVIOR

In Sections 6.1 and 6.2, the end point depended on the environment only through the ability of the system to exchange some substance with it. More generally, the position of the equilibrium or steady-state point may be influenced by the concentrations of certain materials in the environment as well as by factors such as temperature, light, gravitational forces, and pressure. Indeed, one method of studying the rate behavior of a system that is in an equilibrium or "steady state" is to experimentally change the end point and to measure the rate at which the system "relaxes" to the new end point.

It is easy to see how periodic changes in the environment could cause cycling of the steady-state condition for various types of systems. The most obvious of these periodic environmental changes are diurnal and seasonal fluctuations in temperature, light intensity, humidity, and so on.

In systems with linear rate processes, any cycling produced in the system by interaction with a cycling environment has the same frequency as the

6.10 SOME REMARKS ON CYCLING BEHAVIOR

environmental cycling, with some lag and decreased amplitude. There are, however, certain types of systems that interact in a more complex way with the environment, so that the cycling of the system is at a different frequency and possibly with greater amplitude.

Still other types of systems are able to generate their own oscillations in the absence of any environmental cycling. Such systems generally involve feedback relations and are discussed in Chapter 7. When cyclic behavior is observed in a biological system, it is often difficult to determine whether the oscillations are internally generated or whether they are a result of environmental cycling.

CHAPTER
SEVEN
FEEDBACK CONTROL AND STABILITY OF BIOLOGICAL SYSTEMS

Self-control is the quality that
distinguishes the fittest to survive.
George Bernard Shaw

In this chapter we turn attention to the last step in the modeling recipe developed in Chapter 2: investigation of how the overall behavior of the system arises from the properties of the individual dynamic processes and the interactions between them. In particular, we are concerned with those system characteristics that permit the system to maintain a measure of independence from its immediate environment.

Two examples are used as the primary illustrations: the simple public address system of Section 2.1 and the prey–predator ecosystem of Section 6.7. Other examples are discussed in Appendix A.

7.1. STEADY STATES AND EQUILIBRIUM

The concept of the steady state plays a central role in the discussions of this chapter. As defined in Section 6.2, a system is in a steady-state condition when the point that represents the system in state space is stationary with respect to time.

It is important to realize that even if the point representing the whole system is stationary, this does not mean that nothing is going on inside the system. In the two-compartment systems discussed in Sections 6.1 and 6.2,

7.1 STEADY STATES AND EQUILIBRIUM

the transfer processes are in full operation at the steady state. It is the dynamic balance between them that keeps the system stationary. In comparing those two examples, a distinction should be drawn between the equilibrium condition of the isolated system and the steady-state condition of the open system. That distinction is the subject of this section.

Let us suppose that we are looking at a completely stagnant pond. Although all the observable characteristics of the pond (volume, temperature, etc.) may appear to be constant, the individual molecules are bouncing around, changing position, exchanging energy with each other and generally having a wild time. However, if we were to observe any point within the pond, we would see that the number of molecules passing that point per unit time in one direction (actually, we should be speaking of *expected* number) is the same as the number passing in the opposite direction. The net flow past the point is, therefore, zero. This is true for every point in the pond.

The same type of argument might be made relative to any characteristic of the individual units of this system. During any period of time, a number of individual units may change in some direction relative to the given characteristic, but the same number change in exactly the opposite direction. Therefore, the system as a whole remains unchanged; it is said to be at *equilibrium* relative to this characteristic.

The situation is different if there is an inflow and outflow of water, say at opposite ends of the pond. The total amount of water in the pond might still be constant with respect to time. In fact, all the characteristics that were constant in the case of the stagnant pond could just as easily be constant in the new situation. *Something*, however, is different. The "something" is that now there is a net flow-through of water. If we shrink down to the microscopic view again, we see that the amount of water in any region is being held constant because the amount of water flowing out of one side of the region is exactly balanced by the water flowing in on the *other* side of the region. In the stagnant (equilibrium) case, the amount flowing out of one side is balanced by the amount flowing in on the *same* side, so that there is no net flow. Here, however, there is a net flow, which is balanced to produce a *nonequilibrium* steady state.

As a rule, a biological system is not likely to be at equilibrium, though it may be approximately at a nonequilibrium steady state. Suppose, for example, that a lake were at steady state relative to the fish content. This would mean that the number of fish dying would be exactly balanced by the number of fish being born. The number of fish leaving any stage of life through the "top" would be balanced by the number entering from "below." The number of fish would remain constant, but there would be a net flow-through of matter and energy through the component of that lake that we label "fish." By contrast, the lake would be said to be at equilibrium with respect to the fish content only if the number of fish going from one stage of life to the next were balanced by the number returning from the next stage, and the number dying were balanced not by the number being born, but by the number "undying."

Unfortunately, authors are not always careful to distinguish between equilibrium conditions and nonequilibrium steady-state conditions. Confusion between these two terms does no particular harm except that it often results in the misapplication of biophysical and physical—chemical principles that are valid only for equilibrium systems (see Gallucci, 1973, for a discussion).

Many systems that are not at steady state develop in such a way so as to get closer and closer to steady state as time goes on. A number of examples are discussed in Chapter 6. A very simple example that may be useful in building intuition about steady states is provided by a leaky bucket, which is being filled from a water tap (Figure 7.1).

The water comes out of the tap at a constant rate, and some of it escapes from the hole in the bottom. If the input is fast enough, more water will be coming in than going out, and the bucket will begin to fill. As it fills, however, the head pressure above the hole increases, and so the outflow increases. Eventually the rate of outflow increases so that it equals the rate of inflow and the bucket stops filling. With respect to the amount of water in the bucket, the system is now at steady state.

A simple model for this system might involve the following variables:

$$R_{in} = \text{rate of input}$$
$$H = \text{height of water}$$
$$R_{out} = \text{rate of output}$$

These would be related by the symbol–arrow graph shown in Figure 7.2.

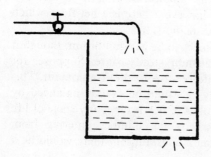

Figure 7.1. Hole-in-the-bucket system.

7.1 STEADY STATES AND EQUILIBRIUM

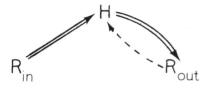

Figure 7.2. Symbol–arrow graph for hole-in-the-bucket system.

This simple system illustrates a number of other things typical of nonequilibrium steady states. First, the amount of water in the bucket at steady state depends on the rate at which the water is coming in from the tap. If the rate from the tap were suddenly decreased, the water level would decrease, thereby decreasing the outflow, until the new outflow would again just balance the inflow. Similarly, the size of the hole has an effect on the steady-state level, since the bigger the hole, the less head is needed for a given outflow. The water tap and the hole in the bottom represent the interactions of the system with the environment. *For a system that tends to approach a steady state, the exact position of the steady-state point in the state space (in this case the height of water in the bucket) is determined by the kind and degree of interaction with the environment.* If the characteristics of the environment are not constant, then the position of the steady-state point is not constant. We may have the kind of situation discussed in subsection 6.1d, in which the system chases a moving end point.

Note also that maintenance of the flows through the system depends on the interactions with the environment. If all interaction with the environment were terminated—the tap shut off and the hole plugged—the level of water would remain constant *at whatever level it had when the interactions were cut off.* After perhaps a few moments of sloshing back and forth, the water would quiet down and there would be no net flow past any point in the system. The system would be at equilibrium.

To see that not every kind of system approaches a steady state, it is only necessary to equip our bucket with a float valve that closes the hole as the level of water goes up. Now the variable graph might look like Figure 7.3. Such a system might continue to fill until it overflows or, if it is fitted with a tight cover, until it bursts like a plasmolysed cell, that is, until the system destroys itself. The difference between the behavior of the systems of Figures 7.2 and 7.3 depends on the differences in the nature of the feedback relations.

Sometimes it is useful (because it is simpler) to limit discussion to descriptions of steady states. In such a context, the signal–flow and symbol–arrow graphs may be simplified. For example, the symbol–arrow

190 FEEDBACK CONTROL AND STABILITY OF BIOLOGICAL SYSTEMS

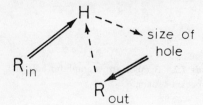

Figure 7.3. Symbol–arrow graph for modified hole-in-the-bucket system.

Figure 7.4. Steady-state symbol–arrow graph for simple prey–predator system.

graph for the prey–predator ecosystem of Appendix A5 (Figure A5.3) simplifies to Figure 7.4, which says the following:

a. A higher light level tends to increase the *steady-state* level of V, the vegetative matter;
b. Higher levels of V tend to increase the *steady-state* levels of P, the prey.
c. Higher levels of P tend to depress the *steady-state* level of V (negative feedback) and increase the steady-state level of Π, the predator.
d. Higher levels of Π tend to depress the *steady-state* level of P (negative feedback).

7.2. STEADY STATES AND SYSTEM STABILITY

The steady state of a system is a kind of balance point in state space. The system is kept there not because of the absense of any force trying to displace it but because of a balance of forces. One might visualize three different types of balance, depending upon what the system does after it is given a momentary "push." The three types of balance are illustrated in Figure 7.5.

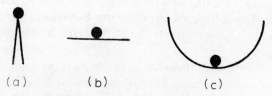

Figure 7.5. Three types of balance point.

7.2 STEADY STATES AND SYSTEM STABILITY

a. Figure 7.5a shows a ball balanced on a knife edge. If the ball is given a momentary push—no matter how small—it takes off for regions unknown. Such a steady state is said to be *unstable*.
b. In the case of a ball resting on a table, any motion is damped by frictional forces. A small push displaces the position of the ball by a small amount. This type of limited stability is often termed *bounded-input–bounded-output stability*.
c. If the ball is contained in the bottom of a trough, then after being displaced, it may bounce back and forth for a while, but it eventually settles back to its original position. This type of stability is termed *asymptotic stability*. The term is meant to convey that if we wait long enough, the system settles back to its original steady state.

One of the first things that one usually tries to do in looking at a system is to determine what the possible steady states are (it may have more than one) and how stable the system is when it is at or near a steady state. By way of example, we take another look at the very simple public address system of Section 2.1. Equation (2.18) for that system was

$$I_M(t) = I(t) + FE_S GE_M I_M(t - \delta) \tag{7.1}$$

where

I_M = net input to the microphone at time t

$I(t)$ = input to the system at time t

F = fraction of output returned to system

E_S = speaker efficiency

G = amplifier gain

E_M = microphone efficiency

δ = system lag factor

This system would be at a steady state if $I_M(t) = I_M(t - \delta)$. So, assuming that this is the case, letting the ss superscript designate the steady-state value and assuming I is kept constant, we get

$$I_M^{ss} = \frac{I}{1 - FE_S GE_M} \tag{7.2}$$

Equation (7.2) shows that a steady state is only possible if $FE_S GE_M < 1$. If this package of parameters is greater than one, then the equation predicts a steady state with a negative value for I_M. This corresponds to no *physically realizable* state. Similarly, if $FE_S GE_M = 1$, there is no physically realizable steady state.

The analysis in Chapter 2 leads to the conclusion that as long as $FE_S GE_M < 1$, the system is stable; if it is not at its steady state, its direction of motion (in state space) is toward the steady state. Note that the steady-state position depends on the input signal. If this is varied, then the system (because of its stability property) will be in the position of having to chase its moving end point (subsection 6.1d).

7.3. STEADY STATES AND OBSERVABILITY

Examining a system at its steady state usually simplifies the mathematics, but generally at the cost of some information (see, for example, the discussion of Marynick and Marynick, 1975). Suppose, for example, that the simple amplifier system were a biological system under study. Without dissecting the system, we might be able to observe I (the system input) and $\mathbb{O}$ (the net system output). Solving the equations of the system for the steady-state relation between input and output,

$$\mathbb{O}^{ss} = \frac{(1-F)E_S GE_M}{1 - FE_S GE_M} I \qquad (7.3)$$

According to this equation, a plot of $\mathbb{O}^{ss}$ versus I is linear. Each point on the plot might be obtained by choosing a value for I and waiting until the system reaches steady state to give a value for $\mathbb{O}^{ss}$. If such a plot is *not* linear, then we know immediately that our description of the system is inadequate. If the relation *is* linear, then we might assert that the experimental results are *consistent* with the model—*not*, of course, that the results prove the model.

If the model is accepted, then a plot of $\mathbb{O}^{ss}$ versus I yields the fraction in equation (7.3) as the slope of the resulting line. Unfortunately, the steady-state observations do not allow us to resolve this package of constants into its components. A variety of strategies have been developed for overcoming the limitations of steady-state measurement without unduly complicating the analysis. Some commonly used strategies are: small displacements from steady state so that "relaxation" back to steady state may be observed; introduction of some marker that allows the observer to distinguish between the "new and old signal," as in tracer and mark–recapture techniques.

7.4. FEEDBACK RELATIONS AND HOMEOSTASIS

An essential aspect of any biological system, be it a subcellular organelle or an entire ecosystem, is the factor of control. No matter how these systems have evolved, certain built-in control features are essential to continued existence. These control networks have the effect of stabilizing the system and allowing it to maintain its own constitution within certain tolerances in the face of varying surrounding conditions. This property has been called the property of homeostasis (homeo = like; static = standing). Striking examples are the constancy of the internal temperature of warm-blooded animals in the face of varying external temperature and the constancy of the pH of the blood in the face of all manner of patent drug preparations.

The property of homeostasis generally involves negative feedback loops, as suggested in Section 2.6. In general, we have a physical variable X, which is determined by another physical variable Y. At the same time, Y is determined by X, but through a different causative relation. That is, we have

$$X = f(Y)$$

$$Y = g(X)$$

but

$$g \neq f^{-1}$$

Now if we look at how X changes as Y is changed, we get

$$\frac{dX}{dY} = \frac{df(Y)}{dY} \qquad (7.4a)$$

whereas

$$\frac{dY}{dX} = \frac{dg(X)}{dX} \qquad (7.4b)$$

In this situation, we have a feedback relation. The feedback is said to be positive or negative according to the sign of the product,[1]

$$\frac{df(Y)}{dY} \cdot \frac{dg(X)}{dX}$$

If the feedback is negative, we have, for example, $Y \Rightarrow X$ and $X \dashrightarrow Y$. An increase in Y leads to an increase in X, which leads to a decrease in Y. Thus changes in the variable Y are resisted. If we look at the "hole in the bucket" system, we see that if the system starts out at the steady state, and the height of the water is forcibly changed (say by dumping in a cup full or dipping out a cup full), the resulting change in the outflow "feeds back" to restore the system to its original state.

Positive feedback, on the other hand, may cause the system to die out or "blow up." In this case, both derivatives are of the same sign, either positive or negative. If they are both positive, then any change in Y causes the same kind of change in X, which causes the same kind of change in Y. That is, the change in Y may feed back upon itself to enhance the originating change. The same is true if both derivatives are negative. In this case, an increase in Y leads to a decrease in X, which leads to an increase in Y. Of course, if the variable Y represents some physical quantity, there generally is a limit on how high the value can go without destroying the system—that is, before the system "blows up." An example of positive feedback on a "biogeophysical" level has been discussed recently by Charney, Stone, and Quirk (1975) in connection with drought in the Sahara. They argue that a decrease in plant cover leads to a change in the radiative characteristics of the ground, which leads to a decrease in rainfall, which leads to a decrease in plant cover.

An excellent nonmathematical discussion of positive and negative feedback loops is contained in the book, *The Limits to Growth* by Meadows et al. (1972).

7.4a. Example: Simple Public Address System

The signal–flow graph (Figure 2.3) shows the feedback loop for the simple public address system we have been considering. The feedback loop can be studied by looking at the relation between any two variables in the loop, say I_M and $\mathcal{O}_S$ (the input to the microphone and the total output from the

---

[1] If the dependencies involve other variables, then partial derivatives may be used. That is, $(\partial f/\partial y)(\partial g/\partial x)$ would describe the feedback loop.

7.4 FEEDBACK RELATIONS AND HOMEOSTASIS

speaker). Equation (2.11) is

$$I_M(t) = I(t) + F\Theta_S(t-\delta) \tag{7.5}$$

Now, suppose that the input I is fixed and constant and that, by some means or other, we fix the value of Θ_S at some constant value. From (7.5) we would find that the resulting value for I_M would be

$$I_M = I + F\Theta_S \tag{7.6a}$$

Next, we solve equations (2.11) through (2.17) for Θ_S in terms of I_M,

$$\Theta_S(t) = E_S I_S(t) = E_S \Theta_S(t) = E_S G I_A(t) = E_S G \Theta_M(t)$$
$$= E_S G E_M I_M(t)$$

Therefore, if somehow we fix I_M (say by adjusting F or I), the resulting value for Θ_S is

$$\Theta_S = E_S G E_M I_M \tag{7.6b}$$

At the steady state, (7.6a) and (7.6b) must be simultaneously satisfied. From (7.6a) and (7.6b), we get

$$\frac{dI_M}{d\Theta_S} = F$$

$$\frac{d\Theta_S}{dI_M} = E_S G E_M$$

These are of the same sign, so apparently the system has a positive feedback loop. The magnitude of the product of the derivatives might be taken as an index of the intensity of the feedback loop. We have already seen that values of this product greater than one lead to the "blow-up" of the system. For values less than one, the effect of the loop may be considered to be sufficiently damped so that the system does not spontaneously blow up.

7.4b. Example: Amplifier Circuit with Negative Feedback

The next step in the development is to modify the circuit so that it has a negative feedback loop and to examine some of the consequences. Let us

196 FEEDBACK CONTROL AND STABILITY OF BIOLOGICAL SYSTEMS

imagine, therefore, that the system is equipped with a device that causes the returning signal to be subtracted from the incoming signal, rather than being added to it. A symbol–arrow graph is shown in Figure 7.6.

The only modification in the set of equations (2.11) through (2.17) is to change a sign in equation (2.11). It becomes

$$I_M(t) = I(t) - F\Theta_S(t-\delta) \qquad (7.7)$$

In place of equation (2.18), we have

$$I_M(t) = I(t) - FE_S GE_M I_M(t-\delta)$$

At the steady state, this gives

$$I_M^{ss} = \frac{I}{1 + FE_S GE_M} \qquad (7.8)$$

which describes a physically realizable steady state for any (positive) values of the parameters.

Going through the same procedure used to get equations (7.6), we get

$$I_M = I - F\Theta_S$$

$$\Theta_S = E_S GE_M I_M$$

and

$$\left(\frac{dI_M}{d\Theta_S}\right)\left(\frac{d\Theta_S}{dI_M}\right) = -FE_S GE_M$$

This is always negative (assuming that the parameters are positive).

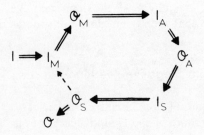

Figure 7.6. Symbol–arrow graph for public address system with negative feedback.

7.4 FEEDBACK RELATIONS AND HOMEOSTASIS

7.4c. Effect of the Feedback on Response to Change in Input

In the previous paragraphs, the point has been made that a negative feedback loop works to oppose disturbances, whereas a positive feedback loop tends to magnify them. In this subsection, we look more explicitly at the sensitivity of the steady-state position to changes in the input variable and how this sensitivity is affected by the feedback loop. The amplifier systems of the two previous subsections are used to illustrate the differing affects of positive and negative feedback.

In these examples, solving for the steady-state value of θ_S (the total speaker output) gives

$$\theta_S^{ss} = \frac{E_s GE_M}{1 - FE_S GE_M} I \quad \text{(positive feedback)}$$

$$\theta_S^{ss} = \frac{E_S GE_M}{1 + FE_S GE_M} I \quad \text{(negative feedback)}$$

The sensitivity of the steady-state speaker output to a change in input can be expressed by the derivative $d\theta_s^{ss}/dI$. To judge the effect of the feedback loop on this sensitivity, we can compare the value of this derivative with the value computed under the assumption that the loop has been interrupted—that is, *opened*. In this example, the loop can be opened by setting F to zero. The result is

$$\frac{\left(\dfrac{d\theta_s^{ss}}{dI}\right)_{\text{open loop}}}{\left(\dfrac{d\theta_s^{ss}}{dI}\right)_{\text{closed loop}}} = \begin{cases} 1 - FE_S GE_M, & \text{positive feedback} \\ 1 + FE_S GE_M, & \text{negative feedback} \end{cases} \quad (7.9)$$

Two things, in particular, should be noted about equation (7.9). The first is that a positive feedback loop increases the sensitivity of the steady-state position to a change of input (ratio less than one), whereas a negative feedback loop decreases the sensitivity (ratio greater than one). That is, the negative feedback loop serves to buffer or insulate the system against changes in the environment (as represented by the value of I). This is especially important for biological systems whose functioning often requires that certain internal state variables be reasonably independent of environmental conditions.

The second thing to note is that in both cases, positive or negative feedback, we have

$$\frac{\left(\dfrac{d\Theta_s^{ss}}{dI}\right)_{\text{open loop}}}{\left(\dfrac{d\Theta_s^{ss}}{dI}\right)_{\text{closed loop}}} = 1 - \frac{dI_M}{d\Theta_s}\frac{d\Theta_s}{dI_M}$$

with the derivatives on the right-hand side taken *around the loop*. This is indicative of a fairly general relation.

The following two systems are compared:

A. Closed loop system.

$$I \rightarrow X \rightleftarrows Y$$
$$X = f(I, Y)$$
$$Y = g(X)$$

B. Open loop system.

$$I \rightarrow X \rightarrow Y$$
$$X = f'(I)$$
$$Y = g(X)$$

with the following relation between the two systems,

$$\frac{df'}{dI} = \frac{\partial f}{\partial I} \tag{7.10}$$

Note that the same rule (the function g) gives the dependence of Y on X in the open and closed loop systems. Different functions must be specified for X, however, since in one case X depends on Y, but in the other case it does not. Nevertheless, the assumption embodied in equation (7.10) is that the direct dependence of X on the input I is the same with or without the feedback loop. Therefore, the relations that are going to be developed are not valid unless this assumption can be made.

An expression for dX/dI in the closed loop system is obtained using the

7.4 FEEDBACK RELATIONS AND HOMEOSTASIS

chain rule (see Appendix B5.5),

$$\frac{dX}{dI} = \frac{\partial f}{\partial I} + \frac{\partial f}{\partial Y}\frac{dY}{dI} \quad (7.11)$$

$$\frac{dY}{dI} = \frac{\partial g}{\partial X}\frac{dX}{dI} \quad (7.12)$$

Substituting (7.12) into (7.11) and solving for dX/dI, gives

$$\left(\frac{dX}{dI}\right)_{\text{closed loop}} = \frac{\partial f/\partial I}{1-(\partial f/\partial Y)(\partial g/;X)} \quad (7.13)$$

For system (B), we simply have

$$\left(\frac{dX}{dI}\right)_{\text{open loop}} = \frac{df'}{dI} \quad (7.14)$$

The ratio of these gives, after algebraic rearrangement and use of assumption (7.10),

$$\frac{(dX/dI)_{\text{open loop}}}{\left(\frac{dX}{dI}\right)_{\text{closed loop}}} = 1 - \frac{\partial f}{\partial Y}\frac{\partial g}{\partial X} \quad (7.15)$$

The derivatives on the right-hand side of (7.15) are taken around the feedback loop. The value of this product of derivatives—or rather, its *negative*—is a measure of the effectiveness of the feedback loop in the stabilization of the steady-state position against changes in the environment. It is termed the *homostatic index* of the loop (abbreviated H.I.); in the literature of system control theory, it is called the *open loop gain*. What we have just found is that under the assumption (7.10)

$$\text{H.I.} = \frac{(dX/dI)_{\text{open loop}}}{(dX/dI)_{\text{closed loop}}} - 1 \quad (7.16)$$

$$= - \text{product of derivatives around the loop} \quad (7.17)$$

Once a model is constructed, the homostatic index may be obtained from equation (7.17). If the feedback loop involves more than two variables, direct application of the chain rule leads to exactly the same prescription: Use the product of the derivatives taken around the loop. That is, if we have

$$X_k \to X_{k-1} \to \cdots \to X_2 \to X_1 \to X_k$$

then

$$\text{H.I.} = -\frac{\partial X_1}{\partial X_2}\frac{\partial X_2}{\partial X_3}\cdots\frac{\partial X_{k-1}}{\partial X_k}\frac{\partial X_k}{\partial X_1}$$

The calculated value of H.I. can be compared with an experimentally determined value, provided it is possible to experimentally interrupt the feedback loop without altering the other functions (see Riggs, 1963, Chapter 5, for a discussion) so that equation (7.15) is valid.

For the two amplifier systems, we have

$$\text{H.I.} = \begin{cases} -FE_S GE_M, & \text{positive feedback} \\ +FE_S GE_M, & \text{negative feedback} \end{cases}$$

These two systems illustrate the following points:

a. H.I. >0 (negative feedback loop). In this case, the feedback loop stabilizes the position of the steady state against changes in the environment.
b. $0 > \text{H.I.} > -1$ (damped positive feedback loop). For H.I. between 0 and -1, the feedback loop makes the steady-state position more sensitive to changes in the environment.
c. H.I. ≤ -1 (strong positive feedback loop). For H.I. equal to or less than -1, the sensitivity may be thought of as being so great as to eliminate the steady state altogether.

Further discussions of interest may be found in von Foerster (1957), Goldman (1960), and in the references of subsections 1.8a and 1.8c.

7.5. EXAMPLE: LOTKA–VOLTERRA MODEL FOR TWO-SPECIES PREY–PREDATOR SYSTEM

In Section 6.7, the following equations were rationalized as a model of direct interaction between two populations:

$$\frac{dN_A}{dt} = \kappa_A N_A + \kappa_{AB} N_A N_B$$

$$\frac{dN_B}{dt} = \kappa_B N_B + \kappa_{BA} N_A N_B$$

7.5 LOTKA—VOLTERRA MODEL

It was pointed out that a variety of kinds of interaction could be described, depending upon the signs of the coefficients. To provide a specific example, we consider a prey–predator system with

P = size of prey population
Π = size of predator population
α_p, α_π = parameters expressing the intrinsic rate of increase of each population in the absence of the other (In particular, α_p may be thought of as including the influence of the environment in the form of food supply and other essentials for the growth of the prey.)
β_p, β_π = parameters expressing the effect of the interaction on each population.

Note that the parameters are taken to be positive, so that the signs of the effects may be explicitly indicated. The equations become

$$\frac{dP}{dt} = \alpha_p P - \beta_p P \Pi \qquad (7.18a)$$

$$\frac{d\Pi}{dt} = -\alpha_\pi \Pi + \beta_\pi P \Pi \qquad (7.18b)$$

As pointed out in subsection 2.4b, the prey and predator subsystems might each be treated as a compartment if the amount of each is measured in terms of biomass, but not if the amounts are measured in numbers of individuals.

The first step in examining the system is to see if there is a steady state. One way to do this is to try to find it. If we find it, then there is one. At the steady state (if there is one), both P and Π become constant. So, we set equations (7.18) equal to zero, and see if we can find values for P and Π that satisfy the resulting equations. That is, if P_0 and Π_0 are the steady-state values for P and Π, then they must satisfy

$$\alpha_p P - \beta_p P \Pi = 0 \qquad (7.19a)$$

$$-\alpha_\pi \Pi + \beta_\pi P \Pi = 0 \qquad (7.19b)$$

Clearly, the point $(\Pi_0, P_0) = (0, 0)$ does the job. This, however, is a totally uninteresting point, since at this point in the state space, there is no system to study. We, therefore, exclude the point $(0, 0)$ and see if there is another steady-state point that we'll call (Π_1, P_1).

Having excluded $P = \Pi = 0$, it is tempting to simply divide (7.19a) by P

and (7.19b) by Π. First, however, we have to check the possibility that *one* of them is zero at the steady state, while the other is not. If $\Pi_1 = 0$ and $P_1 \neq 0$, then equation (7.19a) becomes

$$\alpha_p P_1 = 0$$

Since we are assuming $\alpha_p \neq 0$, this won't do. If $P_1 = 0$ but $\Pi_1 \neq 0$, we get

$$-\alpha_\pi \Pi_1 = 0$$

and this won't do either. Therefore, neither P_1 nor Π_1 can be zero at the steady state we seek.

Now, proceeding to divide (7.19a) by P and (7.19b) by Π and rearranging the result,

$$\Pi_1 = \frac{\alpha_p}{\beta_p} \qquad (7.20a)$$

$$P_1 = \frac{\alpha_\pi}{\beta_\pi} \qquad (7.20b)$$

Since all parameters are positive, this gives a physically acceptable steady state. [If solutions to (7.19) would have required either Π_1 or P_1 to be negative, there might be a solution to the mathematical equations, but one that could not be physically reached.] If P and Π ever take the values given by (7.20), then they will not change (assuming the parameters don't change). In this case, however, if the system is not *at* the steady state, it does not move toward it. In Section 7.7, we look at what it does do.

According to equation (7.17), we have H.I. = 0. However, the conditions for application of (7.15) do not hold, since opening the loop requires that one of the interaction terms be set to zero. This causes *disappearance* of the steady state. That is, it is not only the position, but the very existence of the steady state that depends on the interaction.

Note that equations (7.18) are unrealistic in the sense that they predict that if no predator is present, then the prey increases without bound. Such a model is at best useful within limited regions of the state space; we cannot take seriously any prediction that a physical quantity tends to become infinite as time goes on. This "defect" in the model may be remedied by the inclusion of terms for intraspecies competition, as in Appendix A5. As is pointed out in the appendix, a great many idealizations remain even after the inclusion of such terms.

7.6. SENSITIVITY ANALYSIS

In many feedback systems, it is possible to associate the feedback with some parameters of the system. In such cases, the homeostatic index can be obtained by comparing the response to environmental change when the parameter is zero (open loop) to the response when the parameter is nonzero (closed loop). The "response" that forms the basis of these comparisons is the displacement of the steady-state position. In some cases, as we have seen, no comparison is possible, because no steady state exists when the feedback loop is opened, that is, when the parameter is set to zero. However, it is also possible to look at the effect of small changes in the parameters. Discussion of such effects may be relevant to models such as those of Section 7.5, for which the interpretation of the homeostatic index is rather moot.

It is always important to assess the sensitivity of the behavior of the model to variations in the values of the parameters, since the parameter values can never be known with absolute precision. In many cases, the parameters must be supposed to reflect the influences of the environment. For most models, the parameters should be thought of as random quantities whose expectations appear in the model equations but whose exact values fluctuate in some random manner. In this light, it is obviously important to understand how the behavior of the system changes as the parameter values are varied. Indeed, it may happen that the model indicates an unrealistic sensitivity to such variations, so that the validity of the model is called into question. Some discussions of sensitivity analysis (and further references) may be found in Kowal (1971), Miller (1975), and Miller, Weidhaas, and Hall (1973).

7.7. BEHAVIOR AWAY FROM STEADY STATE

We come back now, to the question of what the system does when it is not at the steady state. The easiest approach usually involves a look at the direction in which the system "moves" in the state space. To illustrate the possibilities, we look at a two-dimensional system consisting of a ball together with a trough, line, or a "two-dimensional knife edge," as in Figure 7.7. We can describe the position of the ball using two coordinates, X and Y. Each circle in the figure is a point at which we might place the ball. The shaded area represents the region in which the ball is physically prevented from being. Through each circle, there is an arrow. It points in the direction that the ball would move in if it were released at that point. The arrow says nothing about the speed of the motion, but only its direction. Let's see whether we can analyze the motion of the ball on the basis of the direction of these arrows. For the moment, we forget about such things as bouncing and shooting past the point of equilibrium. To take these into account, we would have to consider how the motion depends not only on the position of the ball, but also on its momentum. We would have a four-dimensional state space—two position coordinates and two momentum coordinates. Such state spaces are hard to draw.

In Figure 7.7a, if the ball is released at any allowed point that is not on the trough wall, its motion is straight down. Therefore, it must pass through a succession of points, one directly under the other, until it occupies a point on the wall. For points *on* the wall, the motion is still downward, but now begins pointing toward the steady-state point, marked with an asterisk. This point has no arrow through it, because when the system occupies this point, it has no motion—it stays there. The collection of arrows in Figure 7.7a thus shows that if the ball is not at the steady state, its motion in state space is such as to bring it there. Again, nothing is said about how long it might take.

In Figure 7.7b, the points above the line all have directions pointing straight down. A ball released at any of these points passes through a succession of points, one under the other, until it occupies a point on the line. The points on the line have no arrow, since a ball on one of these points goes nowhere. Any of these points is a steady state.

In Figure 7.7c, we see again the situation that a ball released in "free space" has a downward direction. If its downward motion happens to carry it directly to the steady-state point, it gets stuck there (no bouncing, remember). In order for that to happen, it has to be released at a point exactly on the dashed line. If it is released ever so slightly to the left (or the right) of the dashed line, then it winds up very considerably to the left (or

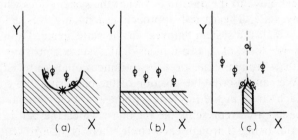

Figure 7.7. Directions of motion for three different kinds of system.

7.7 BEHAVIOR AWAY FROM STEADY STATE

the right). The dashed line separates the state space into regions that are associated with qualitatively different behavior. Such a line is often called a *separatrix* of the system.

Next let's take a slightly more complicated example: a spring with a weight attached to it (Figure 7.8). In this case, we include the effects of overshoots. We can describe the position of the weight by a single coordinate, say in the Y-direction, which also describes the length of the spring. Description of the physical state of the system is completed by giving its velocity (which can be positive or negative) in that direction. The state space then has two dimensions; position (y) and velocity (v). It has an equilibrium position, shown by the asterisk in Figure 7.9 (a, b, or c).

Now, what happens if the spring is stretched to put the weight at point 1 in Figure 7.9? It starts off with zero velocity, but the force of the spring imparts some velocity in the negative direction, so that its motion in state space is described by the arrow at point 1. Having obtained some velocity in the negative direction, the system begins to move. As it approaches the equilibrium length of the spring, the weight acquires greater and greater velocity in the negative direction. Thus as the point moves back toward y^*, it moves toward more negative values of v. When its position is at y^*, its velocity is *not* at $v^* = 0$, but at point 2 in the figure. As the position coordinate gets further away from y^*, the magnitude of the velocity gets less and less, finally reaching $v^* = 0$, when the system is at point 3. The process is repeated, this time with velocity in the positive direction. The velocity reaches a maximum at point 4, with weight overshooting the equilibrium length value.

What happens next? In an ideal system in which no energy is lost by internal friction within the spring, it gets right back to the point from which it started. Since its motion in the state space has been taken as depending only on where it is in the state space, the system has no choice but to repeat the whole thing again and again and again and so on. If

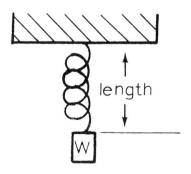

Figure 7.8. Vibrating spring system.

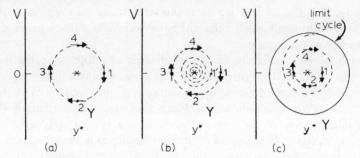

Figure 7.9. State-space diagrams for vibrating spring system.

position were graphed as a function of time, we might get a picture such as in Figure 7.10a; it would continue to oscillate up and down about the steady-state value as the spring itself would be observed to vibrate up and down. Likewise the velocity would oscillate between positive and negative values, as the spring would be observed to speed up, then slow down, then speed up in the opposite direction. If the velocity were plotted against time, the shape of the curve would be similar to that in Figure 7.10a. Of course, the two curves would not be in phase, since one variable reaches an extremum value as the other takes its steady-state value. The dashed curve in Figure 7.9a is a plot of one variable against the other. This curve, the "trajectory in state space," is a closed cycle. In the state-space description, the system cycles about the steady-state point, without ever getting there.

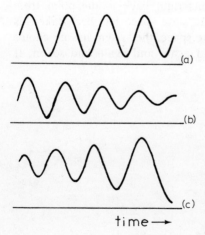

time⟶

Figure 7.10. Oscillating time curves: *a*) sustained oscillations; *b*) damped oscillations: *c*) increasing oscillations.

7.7 BEHAVIOR AWAY FROM STEADY STATE

In an actual spring, some energy would be lost on each cycle. The system would "run out of steam"; that is, its velocity would be reduced to zero *before* the weight got quite back to its original position. Instead, it might only reach point 1' in Figure 7.9b. A plot of position or velocity versus time would show damped oscillations (Figure 7.10b), as we watch the vibrations of the spring die out. In state space (Figure 7.9b) on each succeeding cycle, we see that the system never quite reaches either the velocity or the spatial position that it had on the last one. The system, therefore, spirals down to the steady-state point.

Finally, one might imagine a situation in which the spring system is absorbing energy from an outside energy source.[2] In this case, the spring picks up a little more energy on each succeeding cycle. We get oscillations of *increasing* amplitude (Figure 7.10c), and the state-space picture is that of an *outward spiral*. As has been repeatedly pointed out, physical variables cannot simply increase without bound, so that such a system must eventually change in some way. For example, the spring might break. Another possibility is for the spring to absorb less and less energy on each succeeding cycle, so that it approaches some limit beyond which it does not go. The state-space picture might be something like that shown in Figure 7.9c, in which the system spirals into a *limit cycle*.

In all cases, we must recall that the position of the steady state depends on the specifics of interaction with the environment. The question we are dealing with concerns the behavior of the system when it is not *at* steady state. It may be that some temporary influence has displaced the system from steady state, or that the environmental influences have shifted so that the position of the steady state has changed. The general strategy is to use the equations of the system to determine how the system is changing when some variable, say X, has the value $X^{ss} + \Delta X$.

Directions in state space for the simple prey–predator system of Section 7.5 can be obtained directly from the differential equations (7.18.) Letting P_1 and Π_1 be steady-state values and ΔP and $\Delta \Pi$ be the departures,

$$\frac{dP}{dt} = \alpha_p(P_1 + \Delta P) - \beta_p(P_1 + \Delta P)(\Pi_1 + \Delta \Pi) \quad (7.21a)$$

$$\frac{d\Pi}{dt} = -\alpha_\pi(\Pi_1 + \Delta \Pi) + \beta_\pi(P_1 + \Delta P)(\Pi_1 + \Delta \Pi) \quad (7.21b)$$

---

[2] Such a source could be a vibrating electric motor. The vibration frequency that would allow the spring system to absorb energy would depend upon the mass of the weight and the stiffness of the spring.

Regrouping terms gives

$$\frac{dP}{dt} = (\alpha_p P_1 - \beta_p P_1 \Pi_1) + \alpha_p \Delta P - \beta_p (P_1 \Delta \Pi + \Pi_1 \Delta P + \Delta P \Delta \Pi) \tag{7.22a}$$

$$\frac{d\Pi}{dt} = (-\alpha_\pi \Pi_1 + \beta_\pi P_1 \Pi_1) - \alpha_\pi \Delta \Pi + \beta_\pi (P_1 \Delta \Pi + \Pi_1 \Delta P + \Delta P \Delta \Pi) \tag{7.22b}$$

The first parenthesized term in each of these is zero, so

$$\frac{dP}{dt} = \alpha_p \Delta P - \beta_p (P_1 \Delta \Pi + \Pi_1 \Delta P + \Delta P \Delta \Pi) \tag{7.23a}$$

$$\frac{d\Pi}{dt} = -\alpha_\pi \Delta \Pi + \beta_\pi (P_1 \Delta \Pi + \Pi_1 \Delta P + \Delta P \Delta \Pi) \tag{7.23b}$$

The procedure is now as follows: We pick a point in the state space; that is, we pick a pair of values (Π, P). Then we see from equation (7.23) whether a system that is at that point in state space is increasing or decreasing with respect to P and with respect to Π; that is, we find its direction of motion in the state space. From equations (7.23), it seems plain that the direction must depend on the signs of ΔP and $\Delta \Pi$. Suppose, to keep things simple, we pick a point with $\Delta \Pi = 0$ (points 1 and 3 in Figure 7.11). For such a point, we have

$$\frac{dP}{dt} = \alpha_p \Delta P - \beta_p \Pi_1 \Delta P$$

$$= \alpha_p \Delta P - \beta_p \frac{\alpha_p}{\beta_p} \Delta P$$

$$= 0$$

$$\frac{d\Pi}{dt} = \beta_\pi \Pi_1 \Delta P$$

$$= \frac{\alpha_p \beta_\pi}{\beta_p} \Delta P$$

We find that P is not changing at all, but the change in Π is proportional to ΔP. For $\Delta P > 0$ (point 1), Π is increasing; for $\Delta P < 0$ (point 3), Π is decreasing.

7.7 BEHAVIOR AWAY FROM STEADY STATE

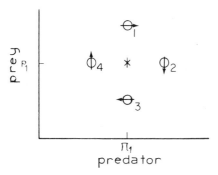

Figure 7.11. Directions in state space for simple prey–predator system.

For $\Delta P = 0$ (points 2 and 4), we find

$$\frac{dP}{dt} = -\beta_p P_1 \Delta \Pi$$

$$= -\frac{\alpha_\pi \beta_p}{\beta_\pi} \Delta \Pi$$

$$\frac{d\Pi}{dt} = -\alpha_\pi \Delta \Pi + \beta_\pi P_1 \Delta \Pi$$

$$= -\alpha_\pi \Delta \Pi + \beta_\pi \cdot \frac{\alpha_\pi}{\beta_\pi} \Delta \Pi$$

$$= 0$$

So for $\Delta P = 0$, Π is not changing, but P is changing in a manner proportional to the negative of $\Delta \Pi$, as shown.

Do these directions make physical sense? Look at the case of Π at its steady-state value ($\Delta \Pi = 0$). If P were also at steady state, then the prey supply would be just enough to sustain the predator population. If P is less than its steady-state value ($\Delta P < 0$), then it is less than that needed to sustain the steady-state level of predators, and the predator population must decrease. The reader should construct analogous arguments for the other points of Figure 7.11.

The directions of the system at these four points certainly do not suggest that the system heads to the steady state no matter where it starts. Indeed, it suggests some type of cycling behavior, such as that shown in Figures 7.9

210   FEEDBACK CONTROL AND STABILITY OF BIOLOGICAL SYSTEMS

and 7.10. We have several possibilities:

a. The system could spiral inward toward the nonzero steady state;
b. The system could spiral outward, "crashing" into the (0,0) steady state;
c. The system could spiral outward and simply execute larger and larger oscillations as time goes on;
d. The system could spiral toward some limit cycle;
e. The system could simply retrace its steps, coming back to its initial starting place, thus executing sustained oscillations whose amplitude is determined completely by where the system starts.

Deciding between these alternatives is generally a difficult mathematical question. The result in this case is that the system behaves according to alternative e.[3]

What does such a result mean? We may imagine that the state space contains a number of possible curves, as shown in Figure 7.12, whose precise shapes depend on the values of the parameters. The initial values of P and Π place the system on one of these curves, and there it stays, cycling around and around. The individual variables continue to oscillate in a manner similar to that portrayed in Figure 7.10.

Comparable analyses of state-space behavior are carried out for somewhat more complex systems in the discussions of Appendix A.

7.8. BIOLOGICAL RHYTHMS

In assessing the significance of the result of the last section, one must again remember that the model itself is a very highly simplified model of a rather complex type of system. Nevertheless, the study of this model by Lotka (1925) and by Voterra (1931) opened up the possibility that cycling behavior, which is frequently observed in biological systems of many types, might arise from the nature of the interactions within the system, rather than simply being a response to periodic stimuli from the environment.

Since the studies of Volterra and of Lotka, there has been a good deal of work on the modeling of systems that show either sustained or damped oscillations. This work has involved the attempt to understand biological rhythms and "biological clock" phenomena in virtually every field of biology—from ecology through physiology and biochemistry. The

---
[3]Mathematical details may be found in Pielou (1966).

7.8 BIOLOGICAL RHYTHMS

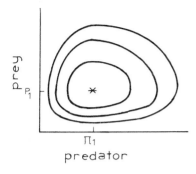

Figure 7.12. Type of curves available as trajectories in the state space of the simple prey–predator system.

frequencies of rhythms that have been studied vary from those expressed in seconds to those expressed in years.

The mathematical descriptions of these various oscillatory systems have certain common characteristics. This, of course, means that certain kinds of relationships between the components of the system are characteristic of oscillatory systems. The understanding of how these physical relations give rise to oscillation requires more background study in mathematics than we are prepared to assume here. Nevertheless, some of the results can be summarized (see Higgins, 1967, for further discussion).

The first result is that oscillations must always be around a steady state; that is, the variables continue to overshoot their steady-state values. The point representing the system in state space winds around the steady-state position.

Of particular importance for the existence of oscillations within the system is the existence of feedback loops, and those feedback loops must exhibit certain characteristics. In particular, if X and Y are two variables whose relationship generates oscillations, then there must be a negative feedback loop connecting the two variables. The feedback loop must involve some sort of time delay, so that the variables do not get to their steady-state values simultaneously (Smith, 1968, discusses this point further). If the system has a limit cycle, then in the immediate vicinity of the steady state one of the variables must influence its own rate of change negatively, while the other influences its own rate of change positively. If the system can be described using differential equations, then the symbol–arrow graph might be of the type shown in Figure 7.13, where the signs of the arrows are determined *at* the steady state. For sustained oscillations of the type shown in Figure 7.12, in which the system is always carried back to its starting value, the derivatives $\partial \dot{X}/\partial X$ and $\partial \dot{Y}/\partial Y$ would

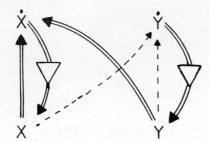

Figure 7.13. One type of symbol—arrow graph that might be associated with an oscillatory limit cycle. Important characteristics are that: i) in the relation between the two variables, there is one negative and one positive feedback: ii) there is one positive and one negative feedback of a variable to its own time derivative.

be zero at the steady state. For example, from equations (7.18), we have

$$\frac{\partial \dot{P}}{\partial P} = \alpha_p P - \beta_p \Pi \qquad (7.24a)$$

$$\frac{\partial \dot{\Pi}}{\partial \Pi} = -\alpha_\pi + \beta_\pi P \qquad (7.24b)$$

Substituting the steady-state values given by expression (7.20) shows that the derivatives in equations (7.24) are zero at the steady state.

These criteria are applied in the discussion of the examples in Appendix A. The relationships are *necessary* but *not sufficient* for oscillations to occur. That means that existence of such relations does not guarantee oscillations, but oscillations cannot occur unless such relations are present. A practical result of this circumstance is that an experimentalist studying the cause of observed oscillatory behavior in any biological system knows that he is looking for feedback loops with certain characteristics.

The behavior of a system undergoing sustained oscillations is stationary in the sense that it is repeating the same pattern over and over. Even though the system itself is not stationary in the state space, its behavior pattern is stationary. Such systems are often referred to as being *stationary in the wide sense*.

EXERCISE 7.1. At this point, the reader should complete the detailed study of at least one of the examples of Appendix A, as well as complete the construction of the model begun in Chapter 2. Particular attention should be paid to the identification of feedback loops and the analysis of their effects upon the behavior of the system.

CHAPTER
EIGHT
CURVE FITTING; ESTIMATING THE PARAMETERS

"With six free parameters
I can draw you
the outline of an elephant.
With a seventh,
I can make it wag its tail."
origin unknown

In the previous chapters, the topic of discussion is the determination of the *form* of the relationship between variables and how this form relates to behavior of the system and the hypothesized mechanisms responsible for that behavior. These relationships involve parameters that can often be related to the expected frequency of occurrence of certain underlying events. This chapter contains some brief comments on the estimation of the values of the parameters from data which is not free of error.

A good deal of the literature of statistics is devoted to this topic. Of the large number of introductory texts in statistics, one might specifically mention Snedecor and Cochran (1967), Steel and Torrie (1960), and Zar (1974). On the topic of curve fitting, the book by Williams (1959) may be suggested as one which contains a good treatment and which is accessible to readers with only moderate mathematical background.

In this chapter, the discussion is limited to relations involving two variables, and which can be represented by some type of smooth curve. The basic concepts apply also to multivariable relations. We let

n = number of data points available

u = number of unknown parameters

8.1. AS MANY PARAMETERS AS DATA POINTS

When the number of data points is exactly equal to the number of parameters, the problem becomes a purely algebraic one. The situation may be summarized as follows:

a. We are given an assumed functional form relating the two variables X and Y,

$$y = f(x; \kappa_1, \ldots, \kappa_u) \tag{8.1}$$

where κ_1 through κ_u are the parameters. For the present purpose, the parameters are the "unknowns."
b. We have $n = u$ data points, which amount to u ordered pairs (x_i, y_i).
c. Equation (8.1) must hold simultaneously for all u data points. That means that we have u separate equations,

$$y_i = f(x_i; \kappa_1, \ldots, \kappa_u); \quad i = 1, \ldots, u \tag{8.2}$$

with u unknowns $\kappa_1, \ldots, \kappa_u$. The set of simultaneous equations is then solved for the parameters $\kappa_1, \kappa_2, \ldots, \kappa_u$.

Such a procedure may also be used to get an equation for a hand-drawn curve, providing the form of the equation is known. Simply read off u pairs of values (x_i, y_i) from the curve, and form the u simultaneous equations as before.

However, a situation in which we determine u parameters from exactly u data points is far from desirable unless we are absolutely certain of the form of the equation *and* that the data points are completely free of error. Let's put it another way. For any set of u data points, we can find many u-parameter equations, each of which conforms exactly to the data (subsection 1.2b). We are left with no room for either compensating for errors in the data or for testing to see whether the right *kind* of equation has been chosen.

To take a simple example, suppose the hypothesized equation is

$$y = \kappa; \quad \kappa \text{ a constant} \tag{8.3}$$

A single data point is enough to determine the value of κ. Few biologists, however, would dream of determining the constant in such a way. Usually κ would be estimated as the mean of at least two observations. Furthermore, if the difference between these two observations is more than could reasonably be ascribed to expected errors, additional observations would have to be made to test the hypothesis that y is constant.

If the hypothesized relation is the straight line,

$$y = a + bx \qquad (8.4)$$

the values of a and b can be determined from any two data points. If the true relation is not a straight line, we will never know it from just two points. Furthermore, even if the true relation *is* a straight line, but one (or both) of the points is appreciably in error, we get the wrong straight line—that is, the wrong values for a and b.

Equations (8.3) and (8.4) are one and two parameter equations, respectively. It should be clear that comparable examples could be given using equations with any number of parameters. For these reasons, one usually likes to have many more data points than parameters. These additional data points provide what are known as *degrees of freedom* that allow the "averaging out" of errors and the testing of the "goodness-of-fit" of the hypothesized functional form. The procedures that need to be used are somewhat more involved than the simple a, b, c recipe already given—just as taking the average of two measurements is more involved than using the value of a single measurement for the case of equation (8.3). Discussion in this chapter is limited to a heuristic examination of concepts involved in the choice of the "best" set of parameters for a given type of equation, when there are some degrees of freedom (more data points than parameters) for estimating the effects of errors.

Before proceeding, it might be well to mention the case of fewer data points than parameters. Suppose, for example, that we have u parameters and $u-1$ data points. From the preceding discussion, we would expect that we could find a set of u parameters that cause the equation to exactly fit these $u-1$ data points together with an arbitrarily chosen uth point. Since there are certainly an infinite number of choices for the uth point, there are an infinite number of sets of parameters that cause the equation to exactly fit the $u-1$ data points. That is, there is not enough data to uniquely determine the parameters.

8.2. WHAT IS "BEST?"

If we have u parameters to estimate from n data points, where $n > u$, the resulting equation does not coincide with all the data points unless the equation expresses the "true" relationship and all the data points are

CURVE FITTING; ESTIMATING THE PARAMETERS

error-free. Since such a situation is never to be expected, we must accept that at least some of the data points will not fall on the "best" curve. Indeed, if we consider equation (8.3), it is clear that the average of two different observed values gives an estimate for κ that coincides with neither, and that the average of three different observed values would have only the merest chance of coinciding with any one of them. In the same way, there is no reason to expect that the "best" curve coincides with any one of the observed data points.

Now the question is how to define "best." An intuitive answer might be that we look for the curve that is "closest" to the data points in some average sense. Such an answer only substitutes one question for another, since we still have to define what we mean by "closest"; but the new question allows a mathematical formulation of the problem. By "closest" one usually means "least distant." This formulation, therefore, leads us to seek a definition of "distance" between the data points and the given equation.

We begin by examining the ordinary ideas of distance between two points, say p_1 and p_2 in Figure 8.1. The distance between the two points is given by the Pythagorean theorem for the hypotenuse of a right triangle, as illustrated in the figure. Recalling that each point can be represented as an ordered pair, say (x_1, y_1) and (x_2, y_2), the distance between the two points can be written

$$d(p_1, p_2) = \left[(x_1 - x_2)^2 + (y_1 - y_2)^2 \right]^{1/2} \qquad (8.5)$$

Note that the distance is *not* the sum of the distances in the two directions, but that the *squared* distance $d^2(p_1, p_2)$ is the sum of the squares

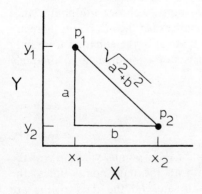

Figure 8.1. Distance between two points is given by the Pythagorean theorem.

8.2. WHAT IS "BEST?"

of the distances

$$d^2(p_1,p_2) = d_x^2 + d_y^2$$

Similarly, if p_1 and p_2 are points in three dimensions, the ordinary distance between them would be

$$d(p_1,p_2) = \left[(x_1-x_2)^2 + (y_1-y_2)^2 + (z_1-z_2)^2\right]^{1/2}$$

or

$$d^2(p_1,p_2) = d_x^2 + d_y^2 + d_z^2$$

Now, let's extend this to the concept of distance in n-dimensions, where n is arbitrary. Suppose we have two points. Each is represented by an ordered n-tuple,

$$p_1 = \left(q_1^{(1)}, q_2^{(1)}, q_3^{(1)}, \ldots, q_n^{(1)}\right)$$

$$p_2 = \left(q_1^{(2)}, q_2^{(2)}, q_3^{(2)}, \ldots, q_n^{(2)}\right)$$

The component of distance between them in the ith direction is $q_i^{(1)} - q_i^{(2)}$. The ordinary (so-called Euclidean) distance[1] between the two points would be expressed by

$$d^2(p_1,p_2) = \sum_{\substack{\text{directions}\\i}} d_i^2 \qquad (8.6)$$

---

[1] This is only one of many possible ways to define a distance, and other definitions may be used in special cases. In general, a *distance* is a function of two points (that is, for any given two points we get a number), with the following special properties:

i. $d(p_1,p_1) = 0$ (the distance from a point to itself is zero).
ii. $d(p_1,p_2) = d(p_2,p_1)$ (the distance is the same in either direction).
iii. Given any three points, $d(p_1,p_3) \leqslant d(p_1,p_2) + d(p_2,p_3)$. (This is called the "triangle inequality." It states that there is no path between any two points p_1 and p_3 that is shorter than the straight line between them. It also states that if two points, say p_1 and p_3 are far apart, then no point, such as p_2, can be close to both of them.)

Any function that satisfies these three requirements qualifies mathematically as a "distance." The function defined by equation (8.7) is the kind of distance most often used. It should be easy to verify that it satisfies i, ii, and iii.

or

$$d(p_1,p_2) = \left[\sum_i \left(q_i^{(1)} - q_i^{(2)} \right)^2 \right]^{1/2} \tag{8.7}$$

Now let us look at our n data points. We have n ordered pairs (x_i, y_i). The total of all the n ordered pairs can be written as a single-ordered $2n$-tuple, $(x_1, y_1, x_2, y_2, \ldots, x_n, y_n)$ corresponding to a point in $2n$-dimensional space. This $2n$-dimensional point represents the data.

We also have an equation with given parameters. For each value of X, the equation can be made to "predict" a value of Y. To keep things distinct, the predicted value of Y corresponding to x_i is denoted $\hat{y}_i$. So, we again can get n ordered pairs $(x_i, \hat{y}_i)$ and an ordered $2n$-tuple $(x_1, \hat{y}_1, x_2, \hat{y}_2, \ldots, x_n, \hat{y}_n)$. This $2n$-dimensional point represents the prediction of the hypothesized relation.

The "distance" between the data and the hypothesized relation may now be taken as being measured by the distance between these $2n$-dimensional points, as given by equation (8.7). Notice, however, that the coordinates corresponding to values of the X variable are the same for both points. As a result, if we assume that all of the measurement error in the data is associated with the Y values, then the distance as given by equation (8.7) is a measure of agreement between hypothesis and data.

Under this assumption, we can look at the distance between the two n-dimensional points,

$$\mathbf{y} = (y_1, y_2, \ldots, y_n)$$

$$\hat{\mathbf{y}} = (\hat{y}_1, \hat{y}_2, \ldots, \hat{y}_n)$$

From equations (8.6) and (8.7), this distance is

$$d(\mathbf{y}, \hat{\mathbf{y}}) = \left[\sum_{i=1}^n d_i^2 \right]^{1/2}$$

$$= \left[\sum_i (y_i - \hat{y}_i)^2 \right]^{1/2} \tag{8.8}$$

The problem now is to find those parameter values that lead to values for the $\hat{y}_i$ that cause $\Sigma(y_i - \hat{y}_i)^2$ to be as low as possible. For obvious reasons, this approach is characterized as a "least squares" technique for estimation of the best parameters.

8.3. THE BEST PARAMETER VALUES FOR LINEAR EQUATIONS 219

To summarize the development so far: For each given value x_i, we have an observed value y_i, resulting in an ordered n-tuple, $\mathbf{y}=(y_1,y_2,\ldots,y_n)$. For each given value x_i, we also have a value $\hat{y}_i$, given by

$$\hat{y}_i = f(x_i; \kappa_1,\ldots,\kappa_u) \qquad (8.9)$$

resulting in the ordered n-tuple $\hat{\mathbf{y}}=(\hat{y}_1,\ldots,\hat{y}_n)$. The choice of the type of function in equation (8.9) may be motivated by mechanistic or correlative considerations. Specific values for the $\kappa_1,\ldots,\kappa_u$ are chosen so as to get the minimum possible value for $d(\mathbf{y},\hat{\mathbf{y}})$ [or, equivalently, for $d^2(\mathbf{y},\hat{\mathbf{y}})$] as given by equation (8.8).

8.3. FINDING THE BEST PARAMETER VALUES FOR LINEAR EQUATIONS

The process of evaluating the parameters to obtain a minimum for $d(\mathbf{y},\hat{\mathbf{y}})$ is carried through for the special case of linear equations,

$$\hat{y}_i = f(x_i; a,b)$$
$$= a + bx_i \qquad (8.10)$$

At the same time, we try to show how the ideas may be extended for use with other types of equations. Readers who are not especially interested in the mathematical details of the process may wish to simply accept expressions (8.20) and (8.21) and skip to Section 8.4.

8.3a. A Useful Characteristic of Extrema

Let us suppose for the moment that we have found the best values for a and b; call them a^* and b^*. If either parameter is made different (greater *or* smaller) than these best values, then $d(\mathbf{y},\hat{\mathbf{y}})$ will increase. The situation is depicted in Figure 8.2. The figure illustrates that the slope of the curve of $d(\mathbf{y},\hat{\mathbf{y}})$ versus a (or of the curve of $d(\mathbf{y},\hat{\mathbf{y}})$ versus b) is zero *at* the best values,

$$\left(\frac{\partial d(\mathbf{y},\hat{\mathbf{y}})}{\partial a}\right)_{a=a^*} = 0 \qquad (8.11\text{a})$$

$$\left(\frac{\partial d(\mathbf{y},\hat{\mathbf{y}})}{\partial b}\right)_{b=b^*} = 0 \qquad (8.11\text{b})$$

220 CURVE FITTING; ESTIMATING THE PARAMETERS

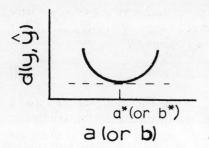

Figure 8.2. $d(y,\hat{y})$ has a minimum value at $a = a^*$ (or $b = b^*$).

Before proceeding, we note that since d must always be positive (at least nonnegative), finding the minimum for $d(y,\hat{y})$ is the same as finding the minimum for $d^2(y,\hat{y})$. Since d^2 is easier to work with, we proceed by finding a and b to minimize $d^2(y,\hat{y})$.

The approach, then, is to find the expressions for derivatives of $d^2(y,\hat{y})$ with respect to a and b, set these equal to zero, and solve for a and b. That is, we look for those values of a and b that give

$$\left(\frac{\partial d^2(y,\hat{y})}{\partial a}\right)_{a=a^*} = 0 \qquad (8.12a)$$

$$\left(\frac{\partial d^2(y,\hat{y})}{\partial b}\right)_{b=b^*} = 0 \qquad (8.12b)$$

In general, if we have a function of any number of variables, say $f(x_1,\ldots,x_n)$, a search for any extremum value would make use of the requirement that all of the $\partial f/\partial x_i$ have to be zero at the extremum. However, in most cases, further examination is required to determine whether the point with zero derivative is an inflection point or an extremum, and if it is an extremum whether it is a minimum or a maximum and whether it is unique (see Figure 8.3).

8.3b. Expressions for Parameters a and b

The derivatives in equations (8.12) can be evaluated using the chain rule (see Appendix B5.5). For parameter b, which is the slope of equation

8.3. THE BEST PARAMETER VALUES FOR LINEAR EQUATIONS

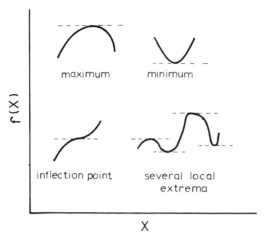

Figure 8.3. Various cases for which df/dx can be zero.

(8.10), we get

$$\frac{\partial d^2(\mathbf{y},\hat{\mathbf{y}})}{\partial b} = \frac{\partial\left(\sum_{i=1}^{n}(y_i-\hat{y}_i)^2\right)}{\partial b}$$

$$= \frac{\partial\left(\sum_{i=1}^{n}(y_i-\hat{y}_i)^2\right)}{\partial(y_i-\hat{y}_i)} \cdot \frac{\partial(y_i-\hat{y}_i)}{\partial b} \quad (8.13)$$

Since

$$\frac{\partial \sum_{i=1}^{n}(y_i-\hat{y}_i)^2}{\partial(y_i-\hat{y}_i)} = \sum_{i=1}^{n}\frac{\partial(y_i-\hat{y}_i)^2}{\partial(y_i-\hat{y}_i)}$$

$$= 2\Sigma(y_i-\hat{y}_i)$$

we get

$$\frac{\partial d^2(\mathbf{y},\hat{\mathbf{y}})}{\partial b} = 2\sum_{i=1}^{n}(y_i-\hat{y}_i)\cdot\frac{\partial(y_i-\hat{y}_i)}{\partial b} \quad (8.14)$$

Now we use the fact that

$$\hat{y}_i = a + bx_i$$

so

$$\frac{\partial (y_i - \hat{y}_i)}{\partial b} = \frac{\partial y_i}{\partial b} - \frac{\partial (bx_i)}{\partial b} - \frac{\partial a}{\partial b}$$

However, since neither y_i nor x_i nor a contain any explicit dependence on b, this is simply

$$\frac{\partial (y_i - \hat{y}_i)}{\partial b} = -x_i$$

Putting this into (8.14),

$$\frac{\partial d^2(\mathbf{y},\hat{\mathbf{y}})}{\partial b} = 2\sum_i (y_i - a - bx_i)(-x_i)$$

$$= 2\sum_i \left(ax + bx_i^2 - x_i y_i\right) \qquad (8.15)$$

Setting (8.15) to zero gives

$$a^*\sum_i x_i + b^*\sum_i x_i^2 - \sum_i x_i y_i = 0 \qquad (8.16)$$

Equation (8.16) has two unknowns, a^* and b^*, so we need another equation before proceeding. It is provided by (8.12a). Again using the chain rule, we wind up with

$$\frac{\partial d^2(\mathbf{y},\hat{\mathbf{y}})}{\partial a} = 2\sum_i (y_i - \hat{y}_i) \cdot \frac{\partial (y_i - \hat{y}_i)}{\partial a} \qquad (8.17)$$

Again, since $\hat{y}_i = a + bx_i$,

$$\frac{\partial (y_i - \hat{y}_i)}{\partial a} = \frac{\partial y_i}{\partial a} - \frac{\partial bx_i}{\partial a} - \frac{\partial a}{\partial a}$$

8.3. THE BEST PARAMETER VALUES FOR LINEAR EQUATIONS

Since y_i, x_i, and b do not depend on a,

$$\frac{\partial (y_i - \hat{y}_i)}{\partial a} = -1$$

Putting this into (8.17),

$$\frac{\partial d^2(\mathbf{y},\hat{\mathbf{y}})}{\partial a} = 2\sum_i (y_i - a - bx_i)(-1)$$

$$= 2\sum_i (a + bx_i - y_i)$$

Setting this equal to zero, gives

$$na^* + b^* \sum_{i=1}^{n} x_i - \sum_{i=1}^{n} y_i = 0 \qquad (8.18)$$

Equations (8.16) and (8.18) are what we need to get a^* and b^*. It makes life simpler, however, to notice that if equation (8.18) is divided through by n, we get

$$a^* + b^*\bar{x} - \bar{y} = 0 \qquad (8.19)$$

where the bar indicates arithmetic average. The rest is a matter of simple algebra. We can, for example, solve (8.19) for $a^* = \bar{y} - b^*\bar{x}$ and substitute into (8.16),

$$(\bar{y} - b^*\bar{x})\sum x_i + b^*\sum x_i^2 - \sum x_i y_i = 0$$

By simple rearrangements, this gives

$$b^* = \frac{\sum_i x_i y_i - \bar{y}\sum_i x_i}{\sum_i x_i^2 - \bar{x}\sum_i x_i} \qquad (8.20)$$

Substituting this expression back into (8.19) and rearranging to solve for a^*,

$$a^* = \frac{\bar{y}\sum_i x_i^2 - \bar{x}\sum_i x_i y_i}{\sum_i x_i^2 - \bar{x}\sum_i x_i} \qquad (8.21)$$

In practice one would simply substitute the numerical value for b^* from (8.20) and solve for a^* directly from (8.19).

The values for a^* and b^* obtained from such a procedure are not to be construed as the true values of the parameters relating y to x. They are only the best (least-squares sense) estimates that we can get for these parameters given the available data, and assuming that all the errors are embodied in y.

8.3c. Graphical Representation

Books in elementary statistics often emphasize the advisability of constructing a graph in which each data point is shown as a point in the $X-Y$ space. If the data is being described by a "best-fitting straight line," the line can be drawn directly on the data plot by calculating the values of two points that lie directly on the line. To do this, we can simply "plug in" any convenient value x_0 into the equation,

$$\hat{y} = a^* + b^* x_0 \qquad (8.22)$$

to get a point $(x_0, \hat{y}_{x_0})$. A second point may be obtained in the same way.

Two points that are especially convenient because they require no additional computation are the points $(\bar{x}, \bar{y})$ and $(-a^*/b^*, 0)$. To verify that these points do lie exactly on the line, we need only plug $\bar{x}$ and $-a^*/b^*$ into (8.22),

$$\hat{y}_{\bar{x}} = a^* + b^* \bar{x}$$

Equation (8.19) already tells us that the right-hand side expression is $\hat{y}_{\bar{x}}$. Similarly, for $x = -a^*/b^*$, we find that $\hat{y} = 0$. Thus the line between the points $(-a^*/b^*, 0)$ and $(\bar{x}, \bar{y})$ is the "least-squares" line of best fit.

8.4. HOW GOOD IS THE BEST FITTING CURVE?

Once we have the parameters that give the smallest $d(\mathbf{y},\hat{\mathbf{y}})$ for a given type of equation, the question is: "Is it small enough to be satisfactory?" If $d(\mathbf{y},\hat{\mathbf{y}})$ is "small," then we are happy. If it is "large," we might consider looking for a different type of equation altogether—possibly with more parameters. The terms "small" and "large" have to be interpreted relative to the size of the expected uncertainties, as discussed in Chapter 4.

Let's put it another way. Suppose we say that for any two data points that we pick, the entire difference between the two Y values can be accounted for in terms of the difference between the two X values, through the equation (8.10). If that were the case, then all the points would lie exactly on the line, and $d(\mathbf{y},\hat{\mathbf{y}})$ would be zero. However, the distance is not zero, because it is not true that all the changes in Y are exactly predictable from the changes in X through the equation. That is, some of the variability in Y is unaccounted for by the equation. This unaccounted for variability can be considered to be of two basic types.

i. The first source is completely random variability that has nothing to do with the change in X. This is the kind of variability that we might see if we measured a lot of Y values at the same value of X. It is usually associated with the uncertainty in the attempt to determine a value of Y for a given value of X.

ii. The second source of error is the failure of the equation to accurately describe the underlying relation between X and Y.

If the value of $d(\mathbf{y},\hat{\mathbf{y}})$ is about what we would expect from source i, then we would have to say that the data are consistent with the type of equation chosen. Although we might wish to use a different type of equation because of *other* information or for theoretical reasons, the observations themselves do not justify a search for a better type of equation. On the other hand, if the value of $d(\mathbf{y},\hat{\mathbf{y}})$ is large compared with what is expected from source i, then we may conclude that the search for a better equation is justified by the data.

In order to compare the value of $d(\mathbf{y},\hat{\mathbf{y}})$ with values expected from source i, it is most convenient to reduce the total distance to an average distance per data point. This would give us something to compare with the error to be expected in a single observation. In taking this "average distance," there are two somewhat strange-looking practices whose reasons I shall try to make clear in a heuristic sort of way.[2]

---

[2]Theoretical justifications may be found in intermediate and advanced texts in statistics.

In arriving at any sort of average, the usual method is to take the total of all the contributed quantities and divide by the number of contributors. In this case, the things that have been added to get a total are not the values of $d(y_i,\hat{y}_i)$, but the values of $d^2(y_i,\hat{y}_i)$. This was because of the additivity properties discussed in Section 8.2. Therefore, it is $d^2(\mathbf{y},\hat{\mathbf{y}})$ that is divided by the "number of contributors."

The second strange-looking practice concerns what is meant by "number of contributors." It has already been pointed out that if we are fitting an equation with u parameters to u data points, we can usually get the value $d(\mathbf{y},\hat{\mathbf{y}})$ to be zero. It is only with the addition of the $(u+1)$st point that we get any contribution to the estimate of error. To divide the error (fitting u parameters with $u+1$ points) by $u+1$ would, therefore, underestimate the error expected in a single observation. As a general rule then, when there are n data points and u parameters to be estimated, we are left with $n-u$ contributors to the error.

Quantities that are often used then, as measures of the "goodness-of-fit" are the *mean square* deviation,

$$MS_{\text{dev.}} = \frac{d^2(\mathbf{y},\hat{\mathbf{y}})}{n-u}$$

$$= \frac{\sum_{i=1}^{n}(y_i-\hat{y}_i)^2}{n-u} \qquad (8.23)$$

and the *root mean square deviation* (also called *standard error of estimate*),

$$RMS_{\text{dev.}} = \left[\frac{d^2(\mathbf{y},\hat{\mathbf{y}})}{n-u}\right]^{1/2}$$

$$= \left[\frac{\sum_{i=1}^{n}(y_i-\hat{y}_i)^2}{n-u}\right]^{1/2} \qquad (8.24)$$

8.5. RANDOM VERSUS SYSTEMATIC DEVIATIONS

If we have a curve that does not exactly fit all the data points, we may characterize the deviations as being either completely random as in Figure 8.4a or systematic as in Figure 8.4b. When the deviations are as in Figure

8.5. RANDOM VERSUS SYSTEMATIC DEVIATIONS

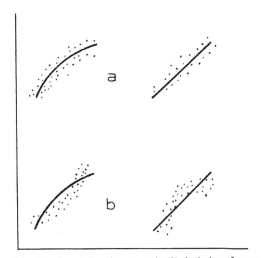

Figure 8.4. Random (*a*) and systematic (*b*) deviations from a curve.

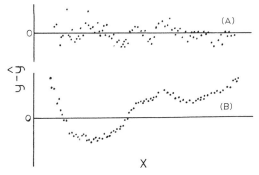

Figure 8.5. Plots of residuals (deviations from a fitted curve) for two different explanatory models for oxygen uptake by tissue slices. Variable y is rate of uptake; x is oxygen concentration in solution: $\hat{y}$ values are computed from the two models.

8.4*a*, it is difficult to believe that source i is not the dominant source of error. On the other hand, when deviations are as in Figure 8.4*b*, it suggests that the contribution from source ii is appreciable, even though the $RMS_{dev.}$ may be small.

In trying to judge whether the deviations are random or systematic, it is often convenient to plot the deviations directly. For each value x_i, we have an observed value of y_i and a calculated value $\hat{y}_i$. The difference $(\hat{y}_i - y_i)$ is called a *residual* value. If all the deviation is from source i, then the residuals will be scattered in some random fashion about zero. Plotting the residuals against values of X often makes it easier to detect systematic trends. Figure 8.5 shows two such plots taken from a study of oxygen uptake of tissue slices as a function of oxygen concentration. The calculated values of $\hat{y}_i$ were based upon two different mechanistic models. It is immediately clear from the *shapes* of these plots that model A is to be preferred.

8.6. RESUMÉ OF THE GENERAL PROCEDURE

Suppose we have an equation with u parameters, so that

$$\hat{y}_i = \hat{y}_i(x_i; \kappa_1, \kappa_2, \ldots, \kappa_u) \qquad (8.25)$$

In order to find the "best" curve, in the sense of minimizing $d^2(y,\hat{y})$ (i.e., the least-squares sense), the procedure to be followed is:

a. Find an expression for each of the derivatives,

$$\frac{\partial \sum_{i=1}^{n} (y_i - \hat{y}_i)^2}{\partial \kappa_j}; \qquad j = 1, \ldots, u$$

b. Set each of these derivatives equal to zero, thereby getting u simultaneous equations;
c. Solve the equations for the u "unknowns" $\kappa_1, \kappa_2, \ldots, \kappa_u$;
d. Test the $MS_{dev.}$ or $RMS_{dev.}$ against expected errors, usually through a statistical-testing procedure;
e. Plot deviations $(\hat{y}_i - y_i)$ versus x_i and examine for pattern.

It should be noted that the calculations involved in step c, especially in the more complicated cases, may require indirect computer algorithms.

8.7. A QUICK AND APPROXIMATE METHOD FOR ESTIMATING A GOOD FITTING CURVE

In this section, we look at an approximate method for estimating the parameters of a relation.[3] The method of approach is to break the data up

---

[3]Interested readers will find discussions of the theoretical basis and range of valid applicability in papers by Wald (1940), Bartlett (1949), Neyman and Scott (1951), and Pavlidis and Horowitz (1974). See also Williams (1959, p. 203).

8.7. A QUICK AND APPROXIMATE METHOD

into groups according to the value of the independent variable (see Figure 8.6), at least as many groups as we have parameters.

If we are trying to fit a straight line, we may divide the data into two groups, as in Figure 8.6a. Next we recall that any straight line must go through the point $(\bar{x},\bar{y})$, where $\bar{x}$ and $\bar{y}$ are determined by the data used. Therefore, if we only use the data in region I, the estimated line would have to go through the point $(\bar{x}_1,\bar{y}_1)$. Similarly, if we only use the data in region II, the line would have to go through $(\bar{x}_2,\bar{y}_2)$. Now if we assume that the same straight line goes through both regions, then the line that connects $(\bar{x}_1,\bar{y}_1)$ with $(\bar{x}_2,\bar{y}_2)$ is an estimate of the line that describes the data.

If the curve that is being used is not a straight line, we may reason as follows: If the data were divided into many small regions, then the curvature within each region would be small. Within such a region, the curve could be approximated by a straight line (see Appendix C). The straight line in region i would, of course, go through $(\bar{x}_i,\bar{y}_i)$. Connecting all of the $(\bar{x}_i,\bar{y}_i)$ then gives an approximation to the curve we seek. To get an approximation to the values of the parameters $\kappa_1,\kappa_2,\ldots,\kappa_u$, we need a minimum of u regions. The relations,

$$\bar{y}_i = f(\bar{x}_i,\kappa_1,\ldots,\kappa_u)$$

then give u simultaneous equations, which may be solved for the parameters.

Choosing more regions than parameters may give some degrees of freedom for testing the adequacy of the equation.

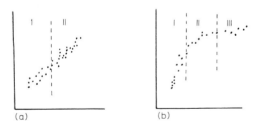

Figure 8.6. Grouping of data points.

8.8. UNEQUAL ERRORS. WEIGHTING OF DATA

The procedure discussed in the previous few sections centers around minimizing the total value of $d^2(\mathbf{y},\hat{\mathbf{y}})$. In this procedure, each data point is considered exactly as important as any other data point. To put it another way, each data point is assumed to be associated with the same amount of measurement error and random variation (source i errors). Therefore, an increment of $d^2(\mathbf{y},\hat{\mathbf{y}})$ of a given size has the same importance regardless of which data point contributes it.

However, it is often the case that our measurements are not all of the same accuracy. In such cases, we do *not* want to count all data points as being equally important. A given increment of $d^2(\mathbf{y},\hat{\mathbf{y}})$ should not be accorded very much importance if it comes from an observation of very doubtful reliability. To take this into account, it is often advisable to multiply each contribution to $d^2(\mathbf{y},\hat{\mathbf{y}})$ by a factor that varies inversely as the *squared* error (since we are summing *squared* deviations) expected for the contributing measurement. These factors are called *weights*. Since the only important consideration is the *relative* magnitude of the weights, they may all be multiplied by the same constant so that they sum to a given convenient quantity.

The procedure for finding the weights might be as follows:

a. For each value of the independent variable x_i, make an estimate of error associated with the measurement of y_i. Call the square of this estimate ε_i^2 (a statistical estimate of the variance of y_i would do beautifully).
b. The weights to be used are proportional to $1/\varepsilon_i^2$;

$$w_i = \frac{c}{\varepsilon_i^2} \qquad (8.26)$$

c. It is often convenient to adjust the weights so that they add up to n, the total number of observations.

This leads to the following series of mathematical statements:

$$n = \sum_{i=1}^{n} w_i$$

$$= \sum_{i=1}^{n} \frac{c}{\varepsilon_i^2}$$

$$= c \sum_{i=1}^{n} \frac{1}{\varepsilon_i^2}$$

8.8. UNEQUAL ERRORS. WEIGHTING OF DATA

Therefore, if we wish the w_i to add to n, each estimated value of $1/\varepsilon_i^2$ is multiplied by

$$c = \frac{n}{\sum_{i=1}^{n} 1/\varepsilon_i^2}$$

If all the ε_i happen to be equal, then this procedure winds up with $w_i = 1$ for all observations. We might say then that a value of $w_i = 1$ is the weight of an "average" observation, whereas $w_i > 1$ or $w_i < 1$ means, respectively, that the observation is accorded more or less than average importance.

For a given vector of weights, $\mathbf{w} = (w_1, w_2, \ldots, w_n)$, the value to be minimized is

$$d_w^2(\mathbf{y}, \hat{\mathbf{y}}) = \sum_{i=1}^{n} w_i (\hat{y}_i - y_i)^2 \qquad (8.27)$$

For $\mathbf{w} = (1, 1, \ldots, 1)$, this is exactly the same as before.

Expression (8.27) is now handled in exactly the same way as when all the weights are one. For the case of a straight line relation, the result is

$$b_w^* = \frac{\sum_i w_i x_i y_i - \bar{y}_w \sum_i w_i x_i}{\sum_i w_i x_i^2 - \bar{x}_w \sum_i w_i x_i} \qquad (8.28)$$

$$a^* = \bar{y}_w - b^* \bar{x}_w$$

where $\bar{x}_w$ and $\bar{y}_w$ are weighted means

$$\bar{x}_w = \frac{1}{n} \sum_i w_i x_i$$

$$\bar{y}_w = \frac{1}{n} \sum_i w_i y_i$$

and

$$\sum_i w_i = n$$

As a practical matter, it is often very difficult to estimate the proper weights to use. When fullest accuracy is desired, the "error structure" of a certain type of data may be the subject of special study. Rough estimates may be made on the basis of the experimenter's experience with the data, the measuring equipment, and so on.

It is interesting to note that when an experienced researcher draws a curve through some plotted data points "by eye," he often automatically "weights" the data. That is, he pays more attention to data points he feels are more accurate and less attention to those he feels are inaccurate. If skillfully done, the result may be a better description of the data than that given by a "least-squares" curve computed without regard to weighting. The difficulty with a curve drawn "by eye" is that it relies not only upon the skill of the performer, but also upon implicit, unexpressed, and often intuitive assumptions about the error structure; assumptions that cannot be held up for scrutiny and critical examination.

Questions concerning the weighting of data may be important when the data are not used in their original form even if all the original measurements are equally accurate. This is because the error structure of the transformed data may be different from that of the original data. This topic is explored in Section 8.9.

8.9. DATA TRANSFORMATION AND THE ERROR STRUCTURE

First of all, why do we ever want to "transform" the data? Usually the situation is that we have an equation that is hard to handle in one form but easy to handle in another form. Straight line equations are usually easiest of all to handle. So, for example, we might wish to use the equation

$$y = e^{kx} \qquad (8.29)$$

in the form

$$\ln y = kx \qquad (8.30)$$

and to plot $\ln y$ versus x. Our concern in this section is how the errors in y relate to the error in $\ln y$.

Another example is equation (6.54),

$$v = \frac{V_{\max} C_s}{K + C_s} \qquad (8.31)$$

8.9. DATA TRANSFORMATION AND THE ERROR STRUCTURE

A commonly used procedure is to write this equation as

$$\frac{1}{v} = \frac{K + C_s}{V_{max} C_s}$$

$$\Rightarrow$$

$$\frac{1}{v} = \frac{K}{V_{max}} \cdot \frac{1}{C_s} + \frac{1}{V_{max}} \tag{8.32}$$

A plot of $1/v$ versus $1/C_s$ should then be a straight line with slope K/V_{max} and intercept $1/V_{max}$. Again, we have the question of how the errors in v and C_s relate to errors in $1/v$ and $1/C_s$.

Of course, equations (8.29) and (8.30) are identical, so it should make no difference which we work with. The same for (8.31) and (8.32). Indeed, it would make no difference if the observations were *error-free*.

Let's look at equation (8.32). To keep things relatively simple, we suppose that C_s can be measured with negligible error and concern ourselves with errors in v. Suppose the true value of v (for a given C_s) is v^*. The result of a measurement differs from the true value by some (unknown) amount ε_v. Letting v be the measured value,

$$v = v^* + \varepsilon_v \tag{8.33}$$

The error ε_v leads to an error $\varepsilon_{1/v}$ in $1/v$, so that

$$\frac{1}{v} = \frac{1}{v^*} + \varepsilon_{1/v}$$

Combining this with (8.33) and solving for $\varepsilon_{1/v}$ in terms of v^* and ε_v gives

$$\frac{-\varepsilon_v}{v^*(v^* + \varepsilon_v)} = \varepsilon_{1/v} \tag{8.34}$$

If we can assume that v^* is much greater than ε_v (otherwise the data are not much good anyhow), we get

$$\varepsilon_{1/v} \cong \frac{-\varepsilon_v}{v^{*2}} \tag{8.35}$$

Equation (8.35) says that if v is measured with constant error over the whole range of the data, then the error in $1/v$ varies at $1/v^2$; if the measurements cover a tenfold range of v, we wind up with $\varepsilon_{1/v}$ covering a range of 100-fold. If v is measured with constant relative error (so that the absolute error ε_v is proportional to v), equation (8.35) tells us that the error in $1/v$ is inversely proportional to v.

The procedure used to arrive at (8.35) can be reformulated to give a general rule: If y is a quantity that is measured with error ε_y, and $f(y)$ is some function of y, then the error in $f(y)$ is approximately (for small errors),

$$\varepsilon_{f(y)} = \left(\frac{\partial f(y)}{\partial y}\right) \cdot \varepsilon_y \qquad (8.36)$$

Equation (8.36) may be understood in the following way. If $f(y)$ and $f(y^*)$ are, respectively, the function values for the measured and the true values of y, then $\varepsilon_{f(y)}$ is $f(y) - f(y^*)$. The derivative is the rate of change of f as y is varied (in the vicinity of y^*); that is, it is an *approximation to* $[f(y) - f(y^*)]/(y - y^*)$, which is the same as $\varepsilon_{f(y)}/\varepsilon_y$. Equation (8.36), therefore, is just an approximate algebraic tautology.

As mentioned at the start of this section, it is often convenient to transform equations to linear form.[4] When this is done, the result of equation (8.36) may be used to estimate the weights discussed in Section 8.8. For further discussion, the reader may wish to consult Brownlee (1965).

8.10. CORRELATIONS BETWEEN VARIABLES

In this section, we look briefly at another way of examining a relation between two variables—more properly, the *correlation* between them.

Change of Ground Rules. In this discussion, we do not need the assumption that all the error lies in the Y direction. Instead, we assume that both variables are measured (neither is preselected) and are subject to error.

Let's begin by looking at two extreme types of situations. The first is the situation in which the "true" value (or expected value) of Y does not at all depend on X. In such a case, the measurements of Y would normally cluster about the mean $\bar{y}$. Deviations from $\bar{y}$ would be random "noise." If we choose to plot the individual values of y_i against the value of x_i that happens to be measured at the same time, we might get a scatter such as in

---

[4]Because of high-speed automatic computers, such transformations are not so necessary as they once were.

8.10. CORRELATIONS BETWEEN VARIABLES

Figure 8.7a. The variations in Y are unrelated to any changes in X; knowing x_i gives no information about the value of y_i. In this case, the randomness or "noise" in Y can be measured by the distance of the vector **y** from a vector, each of whose components is equal to $\bar{y}$,

$$\bar{\mathbf{y}} = (\bar{y}, \bar{y}, \ldots, \bar{y})$$

That is, the randomness would be measured by

$$d^2(\mathbf{y}, \bar{\mathbf{y}}) = \sum_{i=1}^{n} (y_i - \bar{y})^2$$

On the opposite extreme, we might, for example, have a series of solutions of different concentrations of a given chemical compound and study the relation between the logarithm of absorbance of light versus concentration (see subsection 5.5c). Both of these variables can be measured to extremely high accuracy. A plot of one against another might look as in Figure 8.7b. *All* the variation in one variable is meaningful in terms of the other variable. Given one variable, we have essentially total information on the other variable. If the relation is expressed in terms of a hypothetical equation, we find that essentially the entire value of $d^2(\mathbf{y},\bar{\mathbf{y}})$ [or, alternatively of $d^2(\mathbf{x},\bar{\mathbf{x}})$] is accounted for in terms of the hypothesis.

In most real situations of interest, some of the departure of y from $\bar{y}$ is accounted for by the hypothesis, whereas the rest is "noise" (Figure 8.7c). The value of $d^2(\mathbf{y},\bar{\mathbf{y}})$ predicted by the hypothesis might be written as

$$d^2(\hat{\mathbf{y}}, \bar{\mathbf{y}}) = \sum_{i=1}^{n} (\hat{y}_i - \bar{y})^2 \tag{8.37}$$

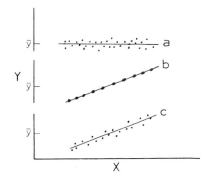

Figure 8.7. Deviations from $\bar{y}$: (*a*) deviations are all "noise"; (*b*) deviations may all be assigned to the relation with x; (*c*) deviations are due partly to noise and partly to relation with x.

In the case represented by Figure 8.7a, this would be zero, whereas for 8.7b, it would be equal to $d^2(y,\bar{y})$. It's useful, then to speak of the *correlation ratio* (or coefficient of determination) of Y given X,

$$\eta^2 = \frac{d^2(\hat{y},\bar{y})}{d^2(y,\bar{y})}$$

$$= \frac{\sum_{i=1}^{n}(\hat{y}_i-\bar{y})^2}{\sum_{i=1}^{n}(y_i-\bar{y})^2} \tag{8.38}$$

as a measure of the dependence of Y on X. It varies from *zero* (as in Figure 8.7a) to *one* (as in Figure 8.7b). It is often referred to as the ratio of explained variation to total variation.

For the special case of a linear relation between Y and X (that is, $y = a + bx$), equation (8.38) becomes the square of the ordinary correlation coefficient,

$$r = \frac{\sum_i (x_i-\bar{x})(y_i-\bar{y})}{\left[\sum_i (x_i-\bar{x})^2 \sum_i (y_i-\bar{y})^2\right]^{1/2}} \tag{8.39}$$

In the form given by equation (8.39), the correlation coefficient has a negative or a positive sign, depending upon the slope of the line relating Y and X.

Anyone who has dealt with sensitive electronic instruments, with high fidelity sound reproduction, and so on, is familiar with the term "signal-to-noise ratio." There is a very close analogy between signal-to-noise ratio and the correlation ratio of equation (8.38). The numerator is a measure of the total "meaningful signal" (i.e., meaningful variability in Y), whereas the denominator is a measure of total variability and would thus be analogous to signal + noise.

8.11. FORCED CORRELATIONS

If you look carefully at equation (8.38), you will see that there are two ways in which the correlation ratio η^2 can be increased—by lowering the

8.11. FORCED CORRELATIONS

denominator or by raising the numerator. It is very easy (unfortunately, it is much *too* easy) to increase η^2 by raising the numerator without increasing the *meaningfulness* of the correlation. To illustrate, we take the case in which Y and X really have no relation. Now, whatever value Y has, just add X to it, and call the result Z,

$$z_i = x_i + y_i$$

The variable Z is going to be a function of X, whether or not Y is. Moreover, the variable Z executes a larger excursion away from its mean than the variable Y does, and *all* of the increased variability is directly accounted for in terms of X. The "signal-to-noise" ratio and the correlation ratio are thereby increased.

Graphically, the situation may be represented as in Figure 8.8. In Figure 8.8a, we see a plot of Y versus X, where

$$y_i = \text{const.} + \varepsilon_i$$

That is, Y is unrelated to X, and the correlation ratio is zero. In Figure 8.8b, we have a plot of Z versus X, where $z_i = y_i + x_i$. The result is that

$$z_i = x_i + \text{const.} + \varepsilon_i$$

The errors are the same for y_i and z_i. Only the "meaningful" variability is changed. We would certainly have to say that Z and X are correlated, whereas Y and X are not. Indeed, we could increase the correlation still further (and get it as close to one as we please), increasing the steepness of the line by letting $z = y + kx$, where k is any number we choose.[5]

Unfortunately, the increased signal-to-noise ratio does not represent any increase in the information content of the relation. It *is*, however, a wonderful tool for misleading the unwary. Indeed, it is easy for a scientist to be trapped into misleading himself. Riggs (1963) cites two ways in which this can happen.

One way is for one of the experimentally measured variables to be a direct, known function of the other variable. In this case, the correlation offers evidence of a relationship, but the relationship could have been predicted without doing any experiments. An example might be the rela-

---

[5]More generally, we might mix the variable y up with x by forming some function

$$z = g(x,y)$$

The error in z (for a given x) can be obtained from the error in y (for that x) by the methods of Section 8.9.

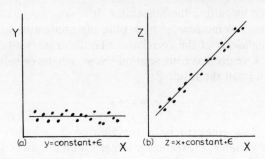

Figure 8.8. The same data plotted according to two different relations illustrates the creation of a valid but meaningless correlation.

tion between total nutrient value of a plant leaf (Y) versus weight of the leaf (X). If a plot of Y versus X gives a straight line, it offers no new information beyond confirming that the relative composition of the leaf remains approximately constant as the leaf grows. On the other hand, if the object is to look for possible changes in relative composition, then a direct measure of concentration should be used, rather than total amount. In real-life situations, the direct relationship between the variables may not be obvious, and may be discovered only after it is looked for. As a general rule, a biologist should be suspicious of any correlation that is *too* good. A correlation that is too good should always be an occasion for asking the question, Is there a direct relation between the variables that has been overlooked?

The second way in which one may force a misleadingly high correlation is by rearranging a hypothetical equation so as to put it into a form that is easy to handle. In such a situation, it is easy to wind up "correlating" one function of x and y with another function of x and y. While such a relationship may be useful in particular cases, one should be cautious about attaching much importance to a high correlation ratio or correlation coefficient arrived at in such a way.

Yet another way in which one may artificially force what appears to be a high correlation may result when no computations at all are made, but the data are simply plotted. The degree of correlation suggested by the plot can be altered by changing the scale of the plot. This is illustrated in Figure 8.9, where the same data has been plotted on three different scales. Dowd and Riggs (1965) refer to the implication of high correlations through a misleading choice of scales as "slight-of-pencil tricks." As a general rule, it is a good idea to choose the scales of the coordinates so that the range of Y observations and the range of X observations are represented by about the same length of paper.

8.11. FORCED CORRELATIONS

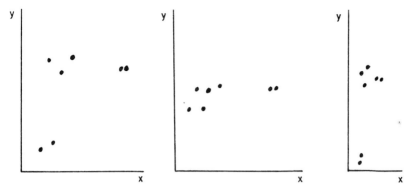

Figure 8.9. The same data plotted on different scale expansions may suggest very different relations between the variables.

EXERCISE 8.1. Assume X to be error-free, and Y to be measured with error ε_y. For each of the following cases, find an expression for ε_z. If x and y both vary over a range of 100-fold, what is the range of ε_z in terms of ε_y?

a. $z = 1/y$
b. $z = xy$
c. $z = y^2$
d. $z = (y^2 + y)x$
e. $z = \ln y$

EXERCISE 8.2. Assume that the following pairs (x_i, y_i) are observations with "no" error in the values of x_i. Plot the points and find *by eye*, the line of best fit.

x_i	1.17	1.74	2.50	3.50	4.00	5.04	7.48	8.00	8.50	9.00
y_i	1.04	1.58	1.83	2.50	3.12	5.08	4.91	7.00	4.80	6.86

EXERCISE 8.3. For the points in Exercise 8.2, find the line of best fit by unweighted least squares. Find the mean square error and correlation coefficient.

EXERCISE 8.4. Assume that the expected error in measuring y is 10% of y. For the data of Exercise 8.2, find the weighted least-squares, best-fitting line.

EXERCISE 8.5. For the data of Exercise 8.2, find the best-fitting line by the method of dividing into regions and finding $(\bar{x}, \bar{y})$ within each region. Two regions will suffice.

CHAPTER NINE
COMPUTING

"And if you take one
from three hundred and sixty-five,
what remains?"
"Three hundred and sixty-four,
of course."
Humpty Dumpty looked doubtful.
"I'd rather see that done
on paper," he said."
Through the Looking Glass
Lewis Carroll

An advertising catch-phrase of the computer industry is "Machines should work, people should think." Nevertheless, it is so easy to get numbers flowing out of a computer that people often wind up using the computer as a substitute for thinking. There is no easy cure for this form of "numeromania."[1] Worse still, while the sufferer may have no difficulty recognizing the disorder in others, he is usually unaware of it in himself. Let the reader beware!

This chapter is concerned with the use of digital computers to produce a set of numbers from a mathematical model. It is not an attempt to teach any one "all about computers" or to explore all the possible uses of computers. It is an attempt to take some of the mystery out of phrases that one sees in the literature, such as "... this result was established by computer simulation." Digital computers can do only a very few simple operations. I hope that the material presented in this chapter gives some

---

[1] Although the medical etiology of this disorder has not been thoroughly investigated, it is believed to be related to the Galatea Delusion discussed by Riggs (1970).

9.1. INSIDE THE MACHINE   241

idea of *what* can be accomplished using these simple operations, *how* it may be accomplished, and what some of the *limitations* are.

9.1. INSIDE THE MACHINE

A digital computer stores numbers and is capable of performing a few simple operations with these numbers. There are many kinds of computers on the market, and there are differences in the specifics of their operation. The description in this section is in terms of a "typical," somewhat primitive computer, but the principles are generally applicable.

In this "typical" computer, the numbers are stored in binary number form encoded as a magnetic or electronic signal. One may conceive of the storage location for a number as consisting of a series of microscopic switches called *bits*. Each of these may be "on" (encoding a *one*) or "off" (encoding a *zero*). An additional switch, or bit, encodes the sign of the number—say, "on" for minus and "off" for plus.

Arithmetic operations are performed at a separate location, which is called a *register* or *accumulator*. In order to add two numbers together, the sequence of operations is as follows: a) the first number is copied from its storage location to the accumulator; b) a copy of the second number is taken from its storage location and added to whatever happens to be in the accumulator (which is the first number); c) the result is copied into a specified storage location.

A programmer working with this primitive computer must express the program in terms of a very few types of elementary operations that it is capable of performing. These elementary operations may be divided into groups. The first group involves only the accumulator:

Instruction Name[2]	What It Does
1. ZERO	Set the accumulator to zero
2. ABS	Whatever is in the accumulator, set the sign bit to "plus"

A second group of operations requires that the computer use a number which is stored at some specified memory location. The memory location is

---

[2]These names are not especially standard. Instructions reside in the computer encoded as numbers. The specific code for each instruction varies from computer to computer.

identified by a number called the *address*. These instructions are

3. ADD   Add the number at the specified address to whatever is in the accumulator

4. SUBTRACT   Subtract the number at the specified address from the number in the accumulator

5. MULTIPLY[3]   Multiply the accumulator contents by the number at the specified address

6. DIVIDE[3]   Divide the number in the accumulator by the number at the specified address

7. FETCH   Simply copy the number at the specified address into the accumulator

8. STORE   Copy the number in the accumulator into the specified address

The instructions themselves are stored in the computer encoded as binary numbers. For instructions of type 3 through 8, one-half of the number encodes the instruction itself, while the other half specifies the required address. There is no way for the computer to know if a particular number stands for an instruction or is intended simply as a number. The computer is equipped with a mechanism, which may be called a "pointer" for finding specified locations. In the ordinary sequence of events, the pointer starts at some particular location or address, and treats the number stored there as an instruction. After completing this instruction, which may involve a number at some specified location, it proceeds to the very next location and again treats the number as an instruction. Thus when the pointer lands at some location it "expects" either a number (if it is in the process of carrying out an instruction) or an instruction (if it is not in the process of carrying out an instruction).

Of course, we do not want the computer to simply continue from one location to the next in this way. For one thing, it would eventually get the computer to the locations in which plain numbers are stored. If the computer gets to such a location expecting an instruction, it "executes the number" if by coincidence the number happens to be the code for a valid

---

[3]More primitive types of computers may not have "multiply" and "divide" as separate instructions. Multiplication and division would be performed as repeated additions or subtractions.

9.1. INSIDE THE MACHINE

instruction. If it is not a valid instruction, then it does whatever it is built to do when it is hopelessly frustrated. For another thing, the real power of the computer lies in its ability to use the result of one computation to "decide" the next operation. Sometimes we wish it to repeat the same set of operations over and over. To accomplish this, two more basic operations may be used:

9. JUMP — Do not execute the next operation in sequence; instead, jump to a specified address and begin executing instructions in sequence from that address on

10. JUMP-IF-POS — Jump to a specified address, but only if the number in the accumulator is greater than zero

These operations also require the specification of an address to jump to.

Finally, we need some instructions that allow numbers to get in and out of the computer:

11. READ — Read a number from some input device, such as a card reader, and copy it into the accumulator;

12. WRITE — Copy the number from the accumulator into some output device, such as an automatic typewriter

A programmer who writes programs in terms of these elementary operations is faced with:

i. Formulating an "algorithm," or a recipe by which the computer may solve the problem using basic arithmetic steps;
ii. Breaking the arithmetic steps into the elementary operations that the computer is capable of executing;
iii. Putting these elementary operations into binary code;
iv. Keeping track of the address of each number used or produced during the computation.

For example, the operation

$$X = A + B$$

would involve the following steps:

a. Determine the address at which A is stored and **FETCH** the number from this address;

b. Determine the address at which B is stored and **ADD** the number from this address;
c. Determine the address at which X is to be stored and **STORE** the accumulator contents at this address.

The operation

$$X = |Y|$$

would involve:

a. **FETCH** from the address of Y
b. **ABS**
c. **STORE** at the address of X.

Now, suppose the computer reads two numbers A and B, and we wish it to write out the greater of the two numbers. That is, we require the computer to make a choice. If $A > B$, then we wish it to print the number A. If $A < B$ then we wish it to print the number B. Furthermore, when it has done this job we wish it to repeat the job with another pair of numbers, and another, and another, and so on. Such a program might be written with twelve instructions. Two locations would be needed to store the numbers A and B. Thus a 14-location computer would be sufficient for the purpose:

Address	Instruction	Address Specified for Instruction
1	READ	Card reader
2	STORE	Location 13 (number A)
3	READ	Card reader
4	STORE	Location 14 (number B)
5	FETCH	Location 13 (number A)
6	SUBTRACT	Location 14 (number B)
7	JUMP-IF-POS	Location 10 (in this case, $A > B$)
8	FETCH	Location 14 (number B, since $A \leq B$)
9	JUMP	Location 11 (get set to write)
10	FETCH	Location 13 (number A, since $A > B$)
11	WRITE	Typewriter
12	JUMP	Location 1 (start over)
13	(stores the number A)	
14	(stores the number B)	

9.1. INSIDE THE MACHINE

Figure 9.1 is a flow chart that diagrams the sequence in which the instructions are to be executed. A useful exercise would be to pick two numbers and carry out the sequence of steps, keeping track of the accumulator contents at the end of each step. Note that if $A = B$, this program will cause the copy stored at B to be printed.

In order to make the use of computers easier, special programs called "compilers" have been developed that convert simple statements such as $X = A + B$ into the required "machine language" instructions. The sequence of steps might be something like the following:

a. The computer user puts a card into a card-punching machine and types $X = A + B$. The machine prints the equation at the top of the card, but it also punches holes in the card. The holes are a code for each symbol. The user reads the type; the computer reads the holes.
b. The computer "reads the holes" and the compiler, which is already in the computer, generates the necessary instructions in terms of the

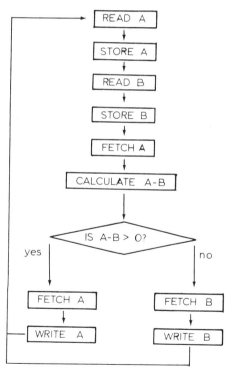

Figure 9.1. Flow chart for program to compute larger of two numbers.

elementary operations. It also sets up the required storage locations for the numbers A, B, and X.

Of course, the user is not free to type in just any kind of statement. It must be a statement of the type that the compiler is equipped to handle. So that the compiler can operate properly, the statements must conform to very strict rules. These rules are sometimes called the grammar or syntax of the language that the programmer uses to communicate with the computer.

There are many different types of compilers in use, each requiring the use of a different set of rules. That is, each accepts a different programming language. Some programming languages, such as COBOL, are intended for business operations and are based on English language words and phrases. Other languages, such as FORTRAN are intended for scientific use and use primarily mathematical symbolism. Some languages, such as PL/1, are intended to be general purpose languages that can be used for a variety of applications. A PL/1 program to print out the larger of two numbers might look like this:

```
GET LIST (A,B);
IF A > B THEN PUT LIST (A);
ELSE PUT LIST (B);
```

There is a very large number of instruction manuals and text books available for each programming language. If you have access to a reasonably large library, you should have little trouble finding one that suits your particular background and taste. However, two that might be mentioned without apology are McCracken (1972) for FORTRAN and Weinberg, Yasukawa, and Marcus (1973) for PL/1.

EXERCISE 9.1 As a mathematical equation, the statement $X = X + 1$ is absurd. As a programming instruction, however, it might make perfectly good sense. What sequence of elementary operation instructions might it involve?

The availability of programming languages allows the computer user to concentrate on part i of the problem—formulating an algorithm that breaks the problem down into basic arithmetic steps.

This discussion should make it clear that contrary to popular myth, computers are not intelligent. On the contrary, they do only what they are told to do, and they need very detailed step-by-step instructions. Their endearing charms come from the fact that they are very faithful and very

9.2. APPROXIMATE CALCULATION OF MATHEMATICAL FUNCTIONS 247

fast. Notice, however, that if you make a mistake in telling it what to do, it makes the mistake very faithfully and *very* fast.
Three conclusions from the previous section are of particular importance.

1. Computation on a digital computer relies essentially on the arithmetic operations of addition, subtraction, multiplication and division, and on comparison of two numbers to see which is larger.
2. In any time period, the computer can only execute a finite number of these operations.
3. The computer only stores finite numbers, and only stores them with finite accuracy. The accuracy depends upon the size of the storage area allocated to each number. If the size of the storage area is enough to store a number with six-place accuracy (in decimal terms), then the computer does not recognize the difference between two numbers that differ only in the seventh place. For example, it will not recognize the difference between 1.0000000 and 0.9999999 or the difference between 0.0000000 and 0.0000001. The size of the storage area can usually be adjusted somewhat, depending upon the accuracy needed, but, of course, it is always finite.

The consequences are that any computation performed on a digital computer must be reduced to a finite number of additions, subtractions, multiplications, divisions, and comparisons, and we must be prepared to accept the consequences of certain rounding errors. The study of this reduction process and the associated errors is called numerical analysis. Introductory treatments may be found in Greenspan (1971), Macon (1963), and Stark (1970).

9.2. APPROXIMATE CALCULATION OF MATHEMATICAL FUNCTIONS

Often, we need to calculate some mathematical function that cannot be expressed in terms of simple arithmetic operations. Some examples are: trigonometric functions such as $\sin\varphi$ and $\cos\varphi$; the exponential function e^x; the logarithmic function $\ln x$; the square root function $\sqrt{x}$. In such a situation, the solution is to find an arithmetic procedure that *approximates* the value needed.

There are many types of strategies that are used in calculating such approximations. This section describes two of the most commonly used strategies.

9.2a. Series Expansions

Many mathematical functions can be expressed by infinite series (see Appendix C). When such series expansions are available, the function value may be caluclated using the elementary arithmetic operations. To get an exact answer, an infinite number of terms would have to be used, and this is not possible. An approximate answer may be obtained, however, by using only the first few terms of the series. If a better approximation is needed, more terms are used.

9.2b. Iterative Procedures

Before discussing the general idea of an iterative procedure, let's look at an example. The example is the calculation of the square root of a number x. We use two properties of the square root of a number. First of all,

$$\frac{x}{x^{1/2}} = x^{1/2} \tag{9.1}$$

Second,

$$\frac{x^{1/2} + x^{1/2}}{2} = x^{1/2} \tag{9.2}$$

As trivial as (9.1) and (9.2) may seem, they are just what's needed. Suppose Q_1 is any number—say it's a guess at $x^{1/2}$. If we divide x by Q_1, we get

$$\frac{x}{Q_1} = R_1 \tag{9.3}$$

If the guess is correct, that is if $Q_1 = x^{1/2}$, then according to (9.1), we wind up with $R_1 = Q_1$. However, if $Q_1 > x^{1/2}$, then $R_1 < x^{1/2}$. Or, if $Q_1 < x^{1/2}$, then $R_1 > x^{1/2}$. Either way, the true value of $x^{1/2}$ must lie between Q_1 and R_1. Equation (9.2) then provides a recipe for making a new guess at $x^{1/2}$,

$$Q_2 = \frac{Q_1 + R_1}{2} \tag{9.4}$$

9.2. APPROXIMATE CALCULATION OF MATHEMATICAL FUNCTIONS

Now, we continue to go back and forth,

$$\frac{x}{Q_2} = R_2$$

$$\frac{Q_2 + R_2}{2} = Q_3$$

and so on. In general, we have

$$\frac{x}{Q_i} = R_i \quad \text{and} \quad \frac{Q_i + R_i}{2} = Q_{i+1}$$

which can be combined into

$$\frac{Q_i + x/Q_i}{2} = Q_{i+1}$$

At each step we get closer and closer to $x^{1/2}$. Since we are interested in finite accuracy only, we quit when $Q_i = Q_{i+1} + \varepsilon$, where the value for ε depends on the accuracy needed.

EXERCISE 9.2. Pick any convenient number, and find its square root by this procedure.

The general idea is that to compute a number whose true value is Q^*, we try to find an *iteration function* f, which operates on a guess, and yields a better guess. More generally, the refined guess Q_{i+1} might depend on several previous guesses,

$$Q_{i+1} = f(Q_i, Q_{i-1}, Q_{i-2}, \ldots)$$

The essential thing is that f must involve only a finite number of arithmetic operations and that successive iterations must give better approximations to Q^*; that is,

$$\lim_{i \to \infty} Q_i = Q^*$$

9.2c. Built-In Functions

Fortunately, the programmer does not usually have to worry about writing programs that compute such standard types of functions as logs, square roots, trigonometric functions, and so on. The programs to compute such functions are usually stored somewhere in the computer. In a FORTRAN program, for example, the statement Q = SQRT(X) would cause the inclusion of the instructions necessary to compute $x^{1/2}$. There are comparable statements for other frequently used functions.

9.3. COMPUTER SIMULATION OF SYSTEMS

For the purpose of simulation, systems may be divided into two types: those in which the state variables vary continuously and those in which changes occur through discrete events. For both types of systems, there are available general purpose simulation programs. To use such a program, the computer user need only specify the structure of the model (in a prescribed form), the starting values of the variables, and the values of the parameters. The general purpose program then completes the details of the simulation. This section discusses the principles on which such simulations are based. For further discussion on both the principles of simulation and on the use of specific simulation programs, the books by Gordon (1969) and by Pritsker (1974) may be suggested.

9.3a. Simulation of Systems with Continuous Dynamics

A system that can be thought of as changing continuously in time is usually described by a set of rate laws or differential equations. Once these differential equations are formulated, we usually want to know the details of the time curve (or the state-space path) for the system under a variety of conditions; that is for different possible values of the parameters and of the initial values. These "predictions" may then be compared with experimental observations for the purpose of determining the "best" values of the parameters (as in Chapter 8) or to test the adequacy of the model. Alternatively, they may be taken as actual predictions of what the system *will* do or what it *would* do under hypothetical conditions.

For relatively simple systems, this information can be obtained by directly solving the differential equations as in Section 5.2. For more complicated systems, computer simulation may be necessary.

Suppose, then, we have a system described by a single differential equation,

$$\frac{dy}{dt} = f(y; \lambda) \tag{9.5}$$

9.3. COMPUTER SIMULATION OF SYSTEMS

where λ stands for the parameters. Recall that the derivative is obtained by taking the limit as $\Delta t \to 0$ of an expression of the type $[y(t+\Delta t) - y(t)]/\Delta t$. It follows (see Appendix C) that equation (9.5) can be replaced by the expression

$$\frac{y(t+\Delta t)-y(t)}{\Delta t} = f[y(t); \lambda] + o(\Delta t) \qquad (9.6)$$

If Δt is small enough, the error will be small, and we can write

$$y(t+\Delta t) \cong y(t) + f[y(t); \lambda]\Delta t \qquad (9.7)$$

Thus if we know $y(t)$, equation (9.7) gives us a way of estimating $y(t+\Delta t)$. Reapplying (9.7) gives an estimate of $y(t+2\Delta t)$. If we start off with $t_0 = 0$ and $y_0 = y(0)$, then repeated application of (9.7) gives

$$y(\Delta t) \cong y_0 + f(y_0; \lambda) \cdot \Delta t$$

$$y(2\Delta t) \cong y(\Delta t) + f(y(\Delta t); \lambda) \cdot \Delta t$$

$$\vdots$$

$$y(k\Delta t) \cong y[(k-1)\Delta t] + f\big(y[(k-1)\Delta t]; \lambda\big) \cdot \Delta t$$

$$\vdots$$

The basic idea is that time is broken up into little "pieces" or intervals, each of length Δt. The derivative is calculated at the start of a given interval. If the derivative is approximately constant throughout the time interval, then multiplying the derivative by the length of the interval gives the approximate change during the interval (equation (9.7)). At the end of the interval, the rate is adjusted by calculating a new value for the derivative.

Of course, an error is introduced by assuming that the derivative remains constant throughout each time interval. The derivative may be changing all the time, which is why it must be adjusted after each time interval. The more often we readjust, the less error is involved. In fact, it can be shown that for the kind of equations we most often deal with, there would be zero error introduced if we adjust infinitely often. But we can't do that. The computer is a finite machine.

Nevertheless, it follows that if the derivative is changing rapidly, we need to adjust more often. That is, the size of the intervals Δt must be smaller. If the derivative is changing only slowly, we can take Δt to be larger and readjust less often. If the derivative doesn't change at all,

$$\frac{dy}{dt} = \text{const.}$$

then we don't have to readjust at all.

Practically all procedures for the computer handling of differential equations use this basic approach. However, many of them are considerably more elaborate in that they use various methods to partially correct for the errors introduced. While such correction methods improve the accuracy for a given choice of Δt, the frequency of "readjustment" is still the limiting factor.

CAUTION. There is a limit to how small Δt can be made. The computer is a finite machine. If Δt is taken too small, the computer won't be able to tell it from zero. As a result, the computed change during the interval Δt is all rounding error.

EXERCISE 9.3. Use this procedure to obtain a graph of y versus t from the equation,

$$\frac{dy}{dt} = -ky; \quad k > 0$$

Use a convenient value for y_0 and a small value for k, say $k = 0.01$. Compute as many points as you think you need to get the "feel" of the procedure. Repeat the procedure with a different choice of Δt. After a certain number of time periods, compare the result with the exact solution, $y(t) = y_0 e^{-kt}$. Observe that if Δt is too large, very peculiar results emerge. Why?

When the system is characterized by several variables, the description of the motion of the system point in state space requires that we describe the change in each of the variables, using a *set* of differential equations. This can be illustrated with the prey–predator system of Appendix A5. Including the vegetation upon which the prey feed, a set of differential equations might be rationalized of the following form (V = vegetation, P = prey,

9.3. COMPUTER SIMULATION OF SYSTEMS

Π = predators),

$$\frac{dV}{dt} = \alpha_v - \beta_v VP \qquad (9.8a)$$

$$\frac{dP}{dt} = \alpha_p VP - \beta_p P\Pi \qquad (9.8b)$$

$$\frac{d\Pi}{dt} = -\alpha_\pi \Pi + \beta_\pi P\Pi \qquad (9.8c)$$

A possible sequence of calculations would be as shown in Figure 9.2.

9.3b. Discrete Event Simulation

As developed in Chapter 5, continuous rate descriptions may be possible when there is a high density of underlying events. In such a case, the

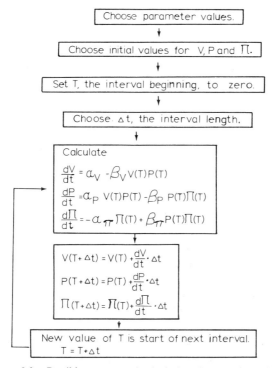

Figure 9.2. Possible sequence of calculations for equations (9.8).

description is in terms of the average rates of these events (that is, expectation of number of events per unit time), which appear as parameters of the differential equations. When the density of events is not high enough, it may be necessary to construct the model directly in terms of the individual events and their probabilities of occurrence.

As a result, continuous models are usually treated as being deterministic. The average rates (i.e., the parameters) do not change, or else change according to a definite pattern. If the system is simulated twice with the same initial conditions and the same initial values for all parameters, identical answers are obtained. On the other hand, discrete event descriptions are usually treated as being stochastic. At each point in the development of the system, there is a choice between alternatives, and the choice is made according to some set of probabilities. If the calculation is done twice, the same answer is not necessarily obtained. However, if the calculation is done many times, we expect to get each possible answer represented. The frequency with which each answer occurs is dictated by the probability distributions that are assumed to obtain within the system.

To accomplish this remarkable feat, a computer program is needed that can generate "random numbers," that is, sequences of numbers with absolutely no discernible pattern. This is not completely possible, but there are programs that produce sequences of numbers that are so nearly unpatterned that they serve nicely as *pseudo*-random number generators.

The way in which such a pseudo-random number generator can be used can be illustrated with the example of Mendel's experiment with pea plants (Chapter 4). We suppose that we have a collection of F_1 plants, and we are about to observe an F_2 plant. According to the model, there are three chances that the F_2 plant will have pink flowers and one chance that it will have white. The observation may be simulated as follows: a) select a number at random, from 1 through 8 (selection at random means that each of these numbers has equal chance of being selected); b) if the selected number is either 1 or 2, then the observation is called "white"; c) if the number is 3 through 8, then the observation is called "pink." It is easy to see that the simulated probabilities are the same as those assumed for the actual system. Furthermore, if the numbers on successive draws are independent of previous draws, then successive draws may be used to simulate successive independent observations. Now, if we repeat the whole thing 929 times, then we have simulated the series of 929 observations discussed in Chapter 4.

Of course, we expect that these 929 simulated observations give neither the exact expectation calculated in Chapter 4, nor exactly the observed outcome. If we wish to know whether the observed outcome is in agree-

9.3. COMPUTER SIMULATION OF SYSTEMS

ment with the hypothesized probability distribution for individual F_2 plants, we might simulate an entire collection of series of observations. That is, we might simulate the series of 929 observations, and then repeat the whole thing say 1000 times. By examining these 1000 repeats, we may be able to make some inference concerning the compatibility of the actual observation with the hypothesis.

The point here is that the simulated results are constructed so that they are *known* to conform to our assumptions. They may, therefore, serve as a kind of experimental control, allowing us to ask the question, Are these assumptions compatible with the observed result? Simulations which involve random sampling within the computer are often called "Monte Carlo" simulations. Note that the experiments can be simulated just as well by throwing dice as by using a pseudo-random number generator; the computer just does it much faster.

The general idea can be extended to quite complicated systems. We might visualize some physical entity, confronted with a whole series of choices. Each choice is made by throwing the dice, or by drawing a random number, according to an assumed probability distribution. When we have gotten to the end, we have one possible "realization" for the behavior of the system. Repeating the whole thing many times gives many "realizations," which are related to each other through the underlying probability distributions that have been assumed. These results may be taken as equivalent to the results of repeated experiments, each of which precisely conforms to our underlying assumptions. Comparing these results with the observed behavior of a real system allows us then to make some inference as to how reasonable the assumptions are for the real system. An example to which such an approach might be applied is the model of bee foraging, discussed in subsection 5.1a.

The example shown in Figure 9.3 is taken from a study in which pest populations are simulated so as to determine appropriate strategies for pest control. The flow chart shows the alternatives available at each stage. The length of time between stages is determined according to an assumed probability pattern, as is the choice at each branch point.

This may seem to put us on somewhat uneasy footing. It is often the case that we just don't have enough information to do more than make a crude guess at these probabilities. This brings us full circle to one of the points made in the beginning chapters. One of the most valuable aspects of the exercise of constructing a model is that in addition to providing a framework for explicit statement of prevailing ideas about the system, it also helps to make the areas of ignorance explicit, and thereby leads to the more intelligent design of new experiments and observations.

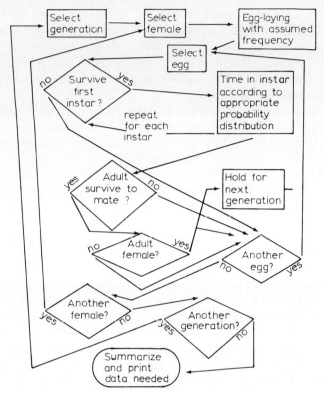

Figure 9.3. Insect population simulation. Adapted from Menke (1974).

9.4. CLASSIFICATION BY COMPUTER; NUMERICAL TAXONOMY

Suppose we were to go into a field and measure the heights of plants at a particular time of the year. Now, suppose that these heights are plotted on a number line, and they fall as shown in Figure 9.4a. No one would have any difficulty in realizing that the plants fall into two distinct groups.

Next, suppose that we measure two characteristics for each plant—height and stem width—and that a plot of one variable against the other looks like Figure 9.4b. In this case also, no one would disagree that the plants fall into two distinct groups. However, in this second case, examination of one variable at a time would not reveal the pattern. Only by the simultaneous consideration of both variables can the grouping pattern be discerned.

What is it about these two patterns that allows their easy recognition? In

9.4. CLASSIFICATION BY COMPUTER; NUMERICAL TAXONOMY   257

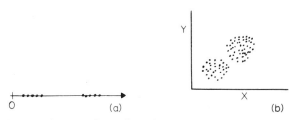

Figure 9.4. Situations in one and two dimensions for which the grouping is obvious.

a general way, the points within each group lie closer to each other than they do to points in the other group. Since the variables represent physical characteristics, points that are closer to each other represent entities that are more similar to each other. Plants within each group are in some way more similar to each other than they are to plants in the other group. In probabilistic terms, two plants picked at random are likely to be more similar if they are picked from the same group than if they are picked from different groups.

Recognition of the grouping or clustering pattern is not always so easy as in the examples of Figure 9.4, for many types of reasons. Some of the reasons are:

a. Number of points may be too large to handle conveniently.
b. Number of characteristics required may make plotting difficult (three dimensions) or impossible (four or more).
c. The clusters may not be so cleanly separated as they are in Figure 9.4, so that the criteria used must be carefully and explicitly formulated. It is not always clear what these criteria should be.
d. Once criteria for cluster separation have been chosen, they may involve considerable calculation.
e. In attempting to recognize patterns of similarity, it is not always clear what characteristics should be examined.
f. The ordinary Euclidian distance between the points (see Section 8.2) might not be the best way to measure the similarity or dissimilarity between the objects being classified. It is not always clear what measure of similarity (or dissimilarity) should be used.
g. The characteristics used for comparison might not be numerical (they might, for example, be color or shape), so that representation as points in a Cartesian space might be of doubtful value.
h. The available data might consist simply of pairwise comparisons, which rate only the similarity between two entities, but do not specify coordi-

nates, which would be required for plotting. This type of situation is often found in psychological studies, but is not limited to psychology. For example, the similarity of the gene complements of two varieties of a plant may be rated on the basis of the properties of the offspring produced by crossing them (see, for example, Murphy, 1974).

NOTE ON TERMINOLOGY. In the literature on taxonomic classification, the objects being classified are referred to as *operational taxonomic units* (OTU's); the characteristics that serve as the basis for the classification are called *characters* (see Sneath and Sokal, 1973, for further discussion of these terms).

Conceptually, items c, e, and f present the most difficulty: the choice of clustering criteria, the choice of characteristics to include in the examination, and the decision as to how similarity should be measured. Once these "policy" decisions have been made, the computer may be of help in their implementation.

Computer clustering procedures involve the following steps:

i. Choose the characters that form the basis of the clustering.
ii. Based on examination of the OTU's, calculate a numerical similarity (or dissimilarity) score for each pair of OTU's. When the characters are numeric, then ordinary distances, computed as in Section 8.2, may be used as dissimilarity scores. For nonnumeric characters, a simple type of similarity score is the fraction of the characters measured for which a particular pair of OTU's agrees. In some studies, as mentioned under item (h), the data themselves are in the form of similarity ratings.
iii. Using the similarity scores, the computer program groups the OTU's.

The criterion used for the actual clustering is usually one of the following types:

i. *Complete linkage*. Complete-linkage clustering requires that once a level of similarity is specified, all members of a cluster must have that degree of similarity (or greater) to *every* other member of the cluster.
ii. *Single linkage*. Single-linkage clustering requires that once a level of similarity is specified, every member of the cluster must have that degree of similarity (or greater) to *at least one* other member of the cluster.
iii. *k-Linkage*. This type of requirement is intermediate between (i) and (ii). Once having specified an integer k and a similarity level, every member of a cluster is required to have that degree of similarity to at least k

9.4. CLASSIFICATION BY COMPUTER; NUMERICAL TAXONOMY

other members of the cluster. For $k = 1$, it is the same as single-linkage clustering.

iv. *Average linkage.* This is another compromise between the extremes of single linkage and complete linkage. It specifies a minimum for the average similarity between a given OTU and the rest of the cluster.

Figure 9.5 illustrates the difference between the two extremes of complete-linkage and single-linkage clustering.

The usual objective is to obtain a partition (see Appendix B1.8) of the set of all OTU's being examined, with the result that OTU's within the same set or cluster are in some way more similar than OTU's in different sets. When the operation is repeated at various levels of similarity, the result may be a hierarchy of nested subsets, which can be represented diagramatically by a type of graph (collection of points or nodes and connections between them) called a *tree*.

In formal language, a tree is a graph that has no loops. The term *tree* also implies that the graph is *connected*; that is, that any node can be reached from any other by "walking" along the connecting lines. Figure 9.6 shows a collection of nodes and connecting lines that is not connected and would be considered to consist of two separate trees.

A simple hypothetical example of the use of trees to represent the result of a hierarchical clustering operation is shown in Figure 9.7. Six OTU's are shown, with a node for each along the top. Any two OTU's that are "very similar" are grouped together in a single cluster; this is represented by connecting the OTU's to a common node, as shown on the line labeled "very similar." The nodes on this line now become the OTU's for the next stage of clustering. The stringency of the similarity requirement is next relaxed, and OTU's that are only "moderately similar" are grouped together, and so on. Even though trees constructed in this way resemble

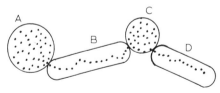

Figure 9.5. Using complete linkage criterion, regions A and C would each be a cluster. By single-linkage criterion, all the points might fall into a single large cluster. By k-linkage criterion with $k = 2$, regions A, B, and C might form a single cluster. But region D would be excluded since the end point does not qualify; therefore, the next point does not qualify, and so on.

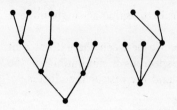

Figure 9.6. A graph consisting of two separate trees.

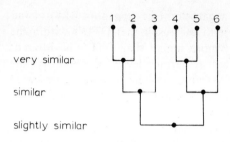

Figure 9.7. Hypothetical example of a phenogram.

phylogenetic trees, they portray phenotypic similarities rather than phylogenetic relationships and are referred to as *phenograms*.

Introductory discussions on numerical taxonomy may be found in Lockhart and Liston (1970) and in Clifford and Stephenson (1975). A more complete treatment may be found in Sneath and Sokal (1973). These books contain numerous references to specific procedures that have been used and to programs that are available. To this list should be added the books by Jardine and Sibson (1971), Anderberg (1973), and Hartigan (1975), which require somewhat more mathematical sophistication.

REFERENCES

"Now, here, you see, it takes all the running *you* can do, to keep in the same place. If you want to get somewhere else, you must run at least twice as fast as that!"
Through the Looking Glass
Lewis Carroll

This list contains references cited in the body of the text. References cited in the appendices are listed at the end of the appropriate sections.

Aleksandrov, A. D., A. N. Kolmogorov, and M. A. Lavrent'ev (Eds.) (1963). *Mathematics: Its Content, Methods and Meaning*, trans. S. H. Gould and T. Bartha, 3 vols., M.I.T. Press, Cambridge, Mass.

Anderberg, M. R. (1973). *Cluster Analysis for Applications*, Academic, New York.

Atkins, G. L. (1969). *Multicompartment Models for Biological Systems*, Methuen, London.

Auslander, D. M., Y. Takahashi, and M. J. Rabins (1974). *Introducing Systems and Control*, McGraw-Hill, New York.

Bailey, N. T. J. (1967). *The Mathematical Approach to Biology and Medicine*, Wiley, New York.

Bartlett, M. S. (1949). "Fitting a Straight Line When Both Variables are Subject to Error," *Biometrics* **5**:207–212.

Batschelet, E. (1975). *Introduction to Mathematics for Life Scientists*, 2nd, ed., Springer-Verlag, New York.

Beckner, M. (1959). *The Biological Way of Thought*, Univ. of California Press, Berkeley.

Bertalanffy, L. von (1968). *General System Theory. Foundations, Development, Applications*, George Braziller, New York.

Brody, B. A. (Ed.) (1970). *Readings in the Philosophy of Science*, Prentice-Hall, Englewood Cliffs, N.J.

Bross, I. D. J. (1972). "Scientific Strategies in Human Affairs: Use of Deep Mathematical Models," *Trans. N. Y. Acad. Sci.* **34**:187–199.

Brownlee, K. A. (1965). *Statistical Theory and Methodology in Science and Engineering*, Wiley, New York.

Charney, J., P. H. Stone, and W. J. Quirk (1975). "Drought in the Sahara: A Biogeophysical Feedback Mechanism," *Science* **187**:434–435.

Clark, L. R., P. W. Geier, R. D. Hughes, and R. F. Morris (1967). *The Ecology of Insect Populations in Theory and Practice*, Methuen, London.

Clifford, H. T., and W. Stephenson (1975). *An Introduction to Numerical Classification*, Academic, New York.

Clow, D. J., and N. S. Urquhart (1974). *Mathematics in Biology. Calculus and Related Topics*, Norton, New York.

Cooper, C. F. (1976). "Ecosystem Models and Environmental Policy," *Simulation* **26**:133–138.

Cowgill, G. R., and D. L. Drabkin (1927). "Determination of a Formula for the Surface Area of the Dog Together with a Consideration of Formulae Available for Other Species," *Amer. J. Physiol.* **81**:36–61.

Crow, J. F., and M. Kimura (1970). *An Introduction to Population Genetics Theory*, Harper and Row, New York.

Curnow, R. N. (1973). "A Smooth Population Response Curve Based on an Abrupt Threshold and Plateau Model for Individuals," *Biometrics* **29**:1–10.

Diamond, J. M., J. Terborgh, R. F. Whitcomb, J. F. Lynch, P. A. Opler, C. S. Robbins, D. S. Simberloff, and L. G. Abele (1976). "Island Biogeography and Conservation: Strategy and Limitations" (An exchange of letters), *Science* **193**:1027–1032.

Dowd, J. E., and D. S. Riggs (1965). "A Comparison of Estimates of Michaelis–Menten Kinetic Constants from Various Linear Transformations," *J. Biol. Chem.* **240**:863–869.

Ellis, B. (1966). *Basic Concepts of Measurement*, Cambridge Univ. Press, Cambridge.

Elsasser, W. M. (1966). *Atom and Organism: A New Approach to Theoretical Biology*. Princeton Univ. Press, Princeton, N.J.

Falconer, D. S. (1960). *Introduction to Quantitative Genetics*, Ronald, New York.

Feller, W. (1957). *An Introduction to Probability Theory and Its Applications*, Vol. I, 2nd ed., Wiley, New York.

Free, J. B. (1970). *Insect Pollination of Crops*, Academic, New York.

Gallucci, V. F. (1973). "On the Principles of Thermodynamics in Ecology," *Ann. Rev. Ecol. Systematics* **4**:329–357.

Garfinkel, D., L. Garfinkel, M. Pring, S. B. Green, and B. Chance (1970). "Computer Application to Biochemical Kinetics," *Ann. Rev. Biochem.* **39**:473–498.

Goel, N. S., and N. Richter-Dyn (1974). *Stochastic Models in Biology*, Academic, New York.

Goldman, S. (1960), "Cybernetic Aspects of Homeostasis," in C. L. Comar and F. Bronner (Eds.), *Mineral Metabolism*, Vol. 1, Part A, Academic, New York.

Gordon, G. (1969). *System Simulation*, Prentice-Hall, Englewood Cliffs, N.J.

Greenspan, D. (1971). *Introduction to Numerical Analysis and Applications*, Markham, Chicago.

Grossman, S. I., and J. E. Turner (1974). *Mathematics for the Biological Sciences*, Macmillan, New York.

Hale, F. J. (1973). *Introduction to Control System Analysis and Design*. Prentice-Hall, Englewood Cliffs, N.J.

REFERENCES 263

Harrison, H. L., O. L. Loucks, J. W. Mitchell, D. F. Parkhurst, C. R. Tracy, D. G. Watts, and V. J. Yannacone (1970). "Systems Studies of DDT Transport," *Science* **170**:503-508.

Hartigan, J. A. (1975). *Clustering Algorithms*, Wiley, New York.

Hemker, H. C., and B. Hess (Eds.) (1972). *Analysis and Simulation of Biochemical Systems*, American Elsevier, New York.

Hempel, C. G. (1965). *Aspects of Scientific Explanation*, Free Press, New York.

Herskowitz, I. H. (1962). *Genetics*, Little, Brown, Boston.

Higgins, J. (1967). "The Theory of Oscillating Reactions," *Ind. Eng. Chem.* **59**:18-62.

Howland, J. L., and C. A. Grabe, Jr. (1972). *A Mathematical Approach to Biology*, Heath, Lexington, Mass.

Hull, D. L. (1974). *Philosophy of Biological Science*. Prentice-Hall, Englewood Cliffs, N.J.

Jacquez, J. A. (1972). *Compartmental Analysis in Biology and Medicine*, Elsevier, New York.

Jardine, N., and R. Sibson (1971). *Mathematical Taxonomy*, Wiley, New York.

Karlin, S. (1975). *A First Course in Stochastic Processes*, 2nd ed., Academic, New York.

Klir, G. J. (Ed.) (1972). *Trends in General System Theory*, Wiley, New York.

Kowal, N. E. (1971). "A Rationale for Modeling Dynamic Ecological Systems," in B. C. Patten (Ed.), *Systems Analysis and Simulation in Ecology*, Vol. 1, Academic, New York.

Kuhn, T. S. (1970). *The Structure of Scientific Revolutions*, 2nd ed., Univ. of Chicago Press, Chicago.

Langhaar, H. L. (1951). *Dimensional Analysis and Theory of Models*, Wiley, New York.

Lewis, W. M. (1976). "Surface/Volume Ratio: Implications for Phytoplankton Morphology," *Science* **192**:885-887.

Lockhart, W. R. and J. Liston (Eds.) (1970). *Methods for Numerical Taxonomy*, Amer. Soc. for Microbiology, Bethesda, Md.

Lotka, A. J. (1925). *Elements of Physical Biology*, Williams & Wilkins, Baltimore (reprinted 1956 as *Elements of Physical Biology*, Dover, New York).

Lucas, H. L. (Ed.) (1962). *The Cullowhee Conference on Training in Biomathematics*, N. C. State Univ., Raleigh.

Macon, N. (1963). *Numerical Analysis*, Wiley, New York.

Marynick, D. S., and M. C. Marynick (1975). "A Mathematical Treatment of Rate Data Obtained in Biological Flow Systems Under Nonsteady State Conditions," *Plant Physiol.* **56**:680-683.

Mather, K. and J. L. Jinks (1971). *Biometrical Genetics*, 2nd ed., Cornell Univ. Press, Ithaca, New York.

McCracken, D. D. (1972). *A Guide to FORTRAN IV Programming*, 2nd ed., Wiley, New York.

McMahon, T. (1973). "Size and Shape in Biology," *Science* **179**:1201-1204.

Meadows, D. H., D. L. Meadows, J. Randers, and W. W. Behrens III (1972). *The Limits to Growth*, New American Library, New York.

Menke, W. W. (1974). "Indentification of Viable Biological Strategies for Pest Management by Simulation Studies," *IEEE Trans. on Systems, Man and Cybernetics*, **SMC-4**:379-386.

Mesarovic, M. D. (Ed.) (1968). *System Theory and Biology*, Springer-Verlag, New York.

Mesarovic, M. D., D. Macko, and Y. Takahara (1970). *Theory of Hierarchical, Multilevel Systems*, Academic, New York.

REFERENCES

Milhorn, T. (1966). *The Application of Control Theory to Physiological Systems*, Saunders, Philadelphia.

Miller, D. R., D. E. Weidhaas, and R. C. Hall (1973). "Parameter Sensitivity in Insect Population Modeling," *J. Theoret. Biol.* **42**:263–274.

Miller, D. R. (1975). "An Experiment in Sensitivity Analysis on an Uncertain Model," *Simulation Today No. 26*, Simulation Councils, Inc., LaJolla, Cal.

Miller, J. G. (1973). "Living Systems. I. The Nature of Living Systems," *Quart. Rev. Biol.* **48**:63–91.

Milsum, J. H. (1966). *Biological Control Systems Analysis*, McGraw-Hill, New York.

Morowitz, H. J. (1967). "Biological Self-Replicating Systems," *Prog. Theoret. Biol.* **1**:35–58.

Murphy, C. F. (1974). "Identification of Diverse Polygenic Systems Affecting Certain Quantitative Expression in Oats," *Tech. Bul.* **223**, Ag. Exp. Station, N. C. State Univ., Raleigh.

Newman, J. R. (Ed.) (1956). *Men and Numbers*, Simon and Schuster, New York.

Neyman, J., and E. L. Scott (1951). "On Certain Methods of Estimating the Linear Structural Relation," *Ann. Math. Stat.* **22**:352–361.

Olby, R. (1974). *The Path to the Double Helix*, Univ. of Wash. Press, Seattle.

Pankhurst, R. C. (1964). *Dimensional Analysis and Scale Factors*, Reinhold, New York.

Pattee, H. H. (Ed.) (1973). *Hierarchy Theory. The Challenge of Complex Systems*, George Braziller, New York.

Patten, B. C. (Ed.) (1971). *Systems Analysis and Simulation in Ecology*, 4 vols., Academic, New York.

Pielou, E. C. (1969). *An Introduction to Mathematical Ecology*, Wiley, New York.

Pollard, E. C. (1969). "The Biological Action of Ionizing Radiation," *Amer. Scientist* **57**:206–236.

Pritsker, A. A. B. (1974). *GASP IV Simulation Language*, Wiley, New York.

Parks, J. S. (1972). "The Physiology of Growth: Animals as Input–Output Devices," Proc. International Summer School on Computers and Research in Animal Nutrition and Veterinary Medicine, Elsinore, Denmark.

Pavlidis, T., and S. L. Horowitz (1974). "Segmentation of Plane Curves," *IEEE Trans. on Computers* **C-23**:860.

Rashevsky, N. (1960). *Mathematical Biophysics. Physico-Mathematical Foundations of Biology*, 2 vols., Dover, New York.

Rensch, B. (1971). *Biophilosophy*. trans. C. A. M. Sym., Columbia Univ. Press, New York.

Rescigno, A., and J. S. Beck (1972). "Compartments," in R. Rosen (Ed.), *Foundations of Mathematical Biology*, vol. II, Academic, New York.

Riggs, D. S. (1963). *The Mathematical Approach to Physiological Problems, A Critical Primer*, Williams & Wilkins, Baltimore (reprinted by M.I.T. Press, Cambridge, Mass., 1970).

Riggs, D. S. (1970). *Control Theory and Physiological Feedback Mechanism*, Williams & Wilkins, Baltimore.

Roberts, F. S. (1976). *Discrete Mathematical Models with Applications to Social, Biological and Environmental Problems*, Prentice-Hall, Englewood Cliffs, N.J.

Rosen, R. (Ed.) (1973). *Foundations of Mathematical Biology*, 3 vols., Academic, New York.

Rubinow, S. I. (1975). *Introduction to Mathematical Biology*, Wiley, New York.

REFERENCES

Russell, B. (1920). *Introduction to Mathematical Philosophy*, 2nd ed., Macmillan, New York.

Schrodinger, E. (1945). *What is Life? The Physical Aspect of the Living Cell*, Macmillan, New York.

Shipley, R. A., and R. E. Clark (1972). *Tracer Methods for In Vivo Kinetics; Theory and Applications*, Academic, New York.

Smeach, S. C., and H. J. Gold (1975). "Stochastic and Deterministic Models for the Kinetic Behavior of Certain Structured Enzyme Systems. II. Consecutive Two-Enzyme Systems," *J. Theoret. Biol.* **51**:79–96.

Smith, J. M. (1968). *Mathematical Ideas in Biology*, Cambridge Univ. Press, Cambridge.

Smith, J. M. (1974). *Models in Ecology*, Cambridge Univ. Press, Cambridge.

Sneath, P. H. A., and R. R. Sokal (1973). *Numerical Taxonomy*, 2nd ed., Freeman, San Francisco.

Snedecor, G. W., and W. G. Cochran (1967). *Statistical Methods Applied to Experiments in Agriculture and Biology*, 6th ed., Iowa State College Press, Ames.

Southwood, T. R. E. (1966). *Ecological Methods, with Particular Reference to the Study of Insect Populations*, Methuen, London.

Stark, P. A. (1970). *Introduction to Numerical Methods*, Macmillan, Toronto.

Steel, R. G. D., and J. H. Torrie (1960). *Principles and Problems of Statistics with Special Reference to Biological Sciences*, McGraw-Hill, New York.

Timofeeff-Ressovsky, N. W., K. G. Zimmer, and M. Delbruck (1935). "Uber die Natur der Genmutation und der Genstruktur," *Nach. Gesel. Wissenchaffen, Gottingen Fachgruppe VI* **1**:190–245.

Titman, D. (1976). "Ecological Competition Between Algae: Experimental Confirmation of Resource-Based Competition," *Science* **192**:463–464.

Volterra, V. (1931). *Lecons sur la Theorie Mathematique de la Lutte Pour la Vie*. Gauthier-Villars, Paris.

von Foerster, H. (1957). "Basic Concepts of Homeostasis," in *Homeostatic Mechanisms*, Brookhaven Symposium in Biology No. 10, 216–242.

Wald, A. (1940). "The Fitting of Straight Lines If Both Variables are Subject to Error," *Ann. Math. Stat.* **11**:285–300.

Watt, K. E. F. (1966). *Systems Analysis in Ecology*, Academic, New York.

Weinberg, G., N. Yasukawa, and R. Marcus (1973). *Structured Programming in PL/C*, Wiley, New York.

Wellington, W. G. (1976). "A Special Light to Steer By," *American Naturalist* **110**:47–52.

Williams, E. J. (1959). *Regression Analysis*, Wiley, New York.

Zar, J. H. (1974). *Biostatistical Analysis*, Prentice-Hall, Englewood Cliffs, N.J.

APPENDIX A
EXAMPLES

The examples discussed in this appendix have been chosen with the following criteria in mind:

a. They are sufficiently complex so as to benefit from a system approach;
b. They may be appreciated, at least at an introductory level, without spending huge amounts of time developing specialized background material;
c. Taken together with the examples discussed in the body of the text, they cover a broad spectrum of biological areas.

The development of each example approximately follows the procedures described in Chapters 2 through 7. However, each type of system requires its own type of approach. It is suggested that the reader look through each of the examples and select one or two for more careful examination.

A1. Regulation of Enzyme Activity
A2. Continuous Culture of Microorganisms
A3. Temperature Control
A4. Regulation of Food Intake and Control of Energy Stores
A5. Simple Prey–Predator Ecosystem

A1. REGULATION OF ENZYME ACTIVITY

A1.1. General Background

From a chemist's point of view, most of the internal mileu associated with a living organism is pretty mild. Temperatures are never extreme and the

A1. REGULATION OF ENZYME ACTIVITY 267

pH is generally near neutrality. In spite of these mild conditions, a living cell is the scene of an enormous amount of chemical activity. This is made possible by enzymes that function as extremely efficient and extremely specific catalysts. Their action is to lower the energy needed to cause a chemical reaction and so increase the speed with which the reaction takes place.

The term *"specific"* is meant to indicate that a particular enzyme only operates on a very limited number of reactions. Regulation of the activity of the individual enzymes therefore becomes a mechanism for the control and regulation of chemical activity.[1] One of the ways in which the activity of some enzymes is regulated is through the action of specific metabolic substances.

The metabolic processes of any living organism comprise a fantastically complex interlocking maze of chemical reactions (see Stadtman, 1970, for a good discussion). One well-studied metabolic process is that of *glycolysis*. This is one of the early stages in the "combustion" of sugar molecules to obtain useable energy. An abbreviated representation of glycolysis might be as shown in Figure A1.1.

Each reaction shown in Figure A1.1 is actually a sequence of several chemical reactions. In this discussion, we focus on the third reaction, which is catalyzed by the enzyme *phosphofructokinase*. We want to de-

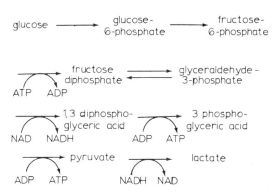

Figure A1.1. Abbreviated glycolysis scheme. Adapted from Higgins, Frenkel, Hulme, Lucas, and Rangazas, 1973. (ATP = adenosine triphosphate; ADP = adenosine diphosphate; NAD and NADH are, respectively, the oxidized and reduced forms of nicotinamide adenine dinucleotide.)

---
[1]Other mechanisms include regulation of amount of enzyme, chemical modification of enzymes, cell compartmentalization so that enzyme and substrate do not have free access to each other.

termine how the production of fructose diphosphate depends upon the amount of fructose-6-phosphate and the total amount of enzyme.

The particular interest in the phosphofructokinase reaction system stems from the fact that it exhibits rather unusual behavior, including the possibility of generating oscillations in the concentration of the reaction product. The enzyme is known to have a complicated chemistry and is affected by several metabolic substances. For the sake of simplicity, we consider the effect of only one of these—fructose diphosphate, which is the product of the reaction. Its effect is to activate the enzyme. The model to be built is based upon the assumption that the enzyme exists in two forms: one active and one inactive. Fructose diphosphate is assumed to combine with the inactive form to produce active enzyme.

A1.2. The Variables, Components, and Compartments

In treating a sequence of chemical reactions, the components of the system may be taken to be the individual reactions that transform the reactants (inputs) to products (outputs). The relevant variables are the concentrations of all the chemical substances (reactants and products) and their time derivatives. The substances important for the system being considered are shown in Table A1.1. We follow the usual practice of designating the concentration of each substance by showing the symbol enclosed in parenthesis and designating the time derivative by showing a dot on top of the symbol.

Table A1.1 Substances Involved in Phosphofructokinase Reaction

Substances	Symbol
fructose-6-phosphate	FP
fructose-1, 6-diphosphate	FDP
active enzyme	E
inactive enzyme	E'
substrate–enzyme complex	X

The dimensions of each of the concentration variables are number of molecules per unit volume. For the time derivatives, the dimensions are number of molecules per unit volume per unit time.

Using the mechanistic formulation suggested in subsection A1.1, the third reaction of Figure A1.1 may be decomposed into the following set of

A1. REGULATION OF ENZYME ACTIVITY

chemical reactions:

$$FP + ATP + E \underset{-1}{\overset{1}{\rightleftarrows}} X \quad (1)$$

$$X \overset{2}{\to} E + FDP + ADP \quad (2)$$

$$FDP + E' \underset{-3}{\overset{3}{\rightleftarrows}} E \quad (3)$$

$$FDP \overset{4}{\to} \text{further reactions} \quad (4)$$

Representation of the system via a component diagram is complicated by the fact that reactions (1) and (3) are reversible. Therefore, the question of which substances are considered to be reactants and which are products depends on the direction in which the system is operating. If both reactions are represented as operating from left to right, then the component diagram of Figure A1.2 results.

From Figure A1.2, it appears that the list of state variables must include (FDP) (which is also an output), (E), (X), and their time derivatives. The outputs may be regarded as (ADP) and (FDP). The inputs which involve FP, ATP, and E' are discussed in the next subsection, when we consider the nature of the dependence of the system on its environment.

Before continuing, it should be pointed out that metabolic sequences are often formulated as compartment models. The specific definition of the compartments depends upon the type of flow being traced. Usually, they are taken to be "pools" of metabolic substances. For example, if the flow of carbon is being traced, Figure A1.1 may be regarded as a compartment diagram. This point of view is particularly valuable in the design and interpretation of tracer experiments.

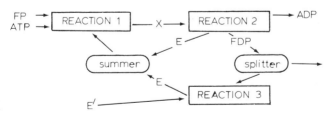

Figure A1.2. Component diagram for phosphofructokinase reaction system. Reactions 1 and 3 are reversible.

A1.3. Interaction with the Environment; Simplifying Assumptions

Before proceeding, we need to make some assumptions regarding the nature of the interaction between the system being studied and its environment.

a. The value of (ATP) is kept constant. Therefore, (ATP) is an input, and $(\dot{\text{ATP}}) = 0$.
b. FP is supplied at a constant *rate*, which we denote by $R_{\text{FP,in}}$. Thus $R_{\text{FP,in}}$ is an input, but the values of (FP) and $(\dot{\text{FP}})$ at any given time depend on other variables as well as on $R_{\text{FP,in}}$. Using the distinction drawn in Chapter 2, (FP) is, therefore, a state variable.
c. The total amount of enzyme in all of its forms (inactive, active, and complexed) is constant, call it E°. Therefore both (E') and its time derivative are state variables (rather than inputs), whose values depend upon the balance between the forward and backward operation of reaction (3). The next assumption concerns this balance.
d. Reaction (3) and its reverse are fast enough so that it may at all times be taken to be at equilibrium. Letting K_3 be the equilibrium constant (Section 6.5),

$$K_3 = \frac{(\text{E})}{(\text{E}')(\text{FDP})}$$

e. Another simplifying assumption, which relates to the internal state of the system, is introduced in the mathematical development of the model, but is stated here for ease of reference. It is that reaction (1) and its reverse are fast enough so that (X) may be considered to be constant. That is, the system is at steady state with respect to (X).

Note that assumptions d and e involve time scale separation, as discussed in Section 6.1. The introduction of these assumptions is motivated by the desire to simplify the mathematical development. The *justification* for the assumptions is that they appear to reasonably reflect the characteristics of the system.

The affect of an assumption like that of d is often difficult to appreciate. The assumption is that the relaxation time (Section 5.4) for (X) is fast compared with that of (FP) or (FDP), so that on a time scale appropriate for discussing (FP) and (FDP), the relaxation time for (X) is effectively zero. It follows that for any *given* value of (FP) and (FDP) the value of (X)

A1. REGULATION OF ENZYME ACTIVITY

may be taken to be independent of how long those values of (FP) and (FDP) have been held. The result is that we drop the variable $(\dot{X})$, and show (X) directly as a function of (FP) and (FDP).

Because of the combination of assumptions c, d, and e, comparable statements may be made concerning (E) and (E'). That is, their time derivatives do not appear as state variables, and any affects of other state variables on them is taken to be instantaneous.

A1.4. Pattern of Relationships Between the Variables

A symbol–arrow graph may be drawn for this system directly from inspection of the chemical reactions, together with a consideration of assumptions a through e and the law of mass action. There is the problem, however, of how to represent assumption c most conveniently. One way is to note that by assumption d, the ratio (E)/(E') is equal to K_3 (FDP). For convenience, therefore, we define a new variable, $K'_3 = K_3$ (FDP) = (E)/(E'), and use it in the symbol–arrow graph (Figure A1.3) instead of (E').

Initially it was stated that our purpose is to determine a relation between FDP production and the concentrations (FP) and E°. Since E° is taken to be constant, it will be of interest to construct a symbol–arrow graph showing the relation between (FP), (FDP), and their derivatives. The resulting abbreviated symbol–arrow graph is shown in Figure A1.4. From Section 7.8, it is seen that the characteristics of this graph suggest the possibility of oscillatory behavior.

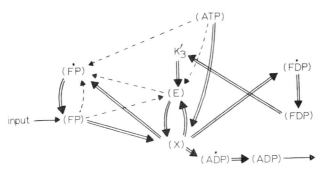

Figure A1.3. Symbol–arrow graph for the phosphofructokinase reaction system: $K'_3 = K_3(\text{FDP})$.

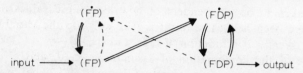

Figure A1.4. Symbol–arrow graph for relation between (FP) and (FDP).

A1.5. The Model Equations

The *net* rate of production for any of the substances is simply its rate of production minus its rate of destruction. Letting v_i be the rate of the ith reaction, v_{-i} be the rate of its reverse, and using the results of Section 5.7, we obtain

$$(\dot{FP}) = R_{FP,in} + v_{-1} - v_1 \tag{A1.1}$$

$$(\dot{FDP}) = v_2 + v_{-3} - v_3 - v_4 \tag{A1.2}$$

$$(\dot{X}) = v_1 - v_{-1} - v_2 \tag{A1.3}$$

Because of assumptions c, d, and e, we do not need explicit equations for $(\dot{E})$ or $(\dot{E}')$. The mathematical form for each of the v_i and v_{-i} is determined by application of the law of mass action,[2] with k_i and k_{-i} denoting the forward and reverse rate constants for the ith reaction. The equation for FDP is simplified by assumption d, which requires $v_3 = v_{-3}$. The results are

$$(\dot{FP}) = R_{FP,in} + k_{-1}(X) - k_1(ATP)(E)(FP) \tag{A1.4}$$

$$(\dot{FDP}) = k_2(X) - k_4(FDP) \tag{A1.5}$$

$$(\dot{X}) = k_1(ATP)(E)(FP) - (k_{-1} + k_2)(X) \tag{A1.6}$$

At this point, use of assumption e allows us to solve (A1.6) for (X),

$$(X) = \frac{k_1(ATP)(E)(FP)}{k_2 + k_{-1}}$$

Assumptions c and e next allow us to eliminate (E) by algebraic substitu-

---

[2]This brings with it its own set of assumptions (see Chapter 5).

A1. REGULATION OF ENZYME ACTIVITY

tion and, after a certain amount of algebraic rearrangement, to get

$$(X) = \frac{k_1 K_3 (\text{ATP})(\text{FP})(\text{FDP}) E^\circ}{[K_3 (\text{FDP}) + 1](k_2 + k_{-1}) + k_1 K_3 (\text{ATP})(\text{FP})(\text{FDP})} \quad (A1.7)$$

Equations (A1.4), (A1.5), and (A1.7) now constitute the equations of the model. The assumptions have allowed us to eliminate a number of state variables. Those that remain are (FP), (FDP), their time derivatives, and (X).

A1.6. Steady States

Since $(\dot{X})$ is always zero by assumption, a steady state for the system is defined by the conditions that $(\dot{\text{FP}}) = 0$ and $(\dot{\text{FDP}}) = 0$. The state in which all concentrations are zero is such a steady state but is uninteresting. We proceed under the assumption that none of the concentrations are identically zero, and look for the existence of a steady state in the (FP)–(FDP) space (Figure A1.5).

Solving equation (A1.4) for $(\dot{\text{FP}}) = 0$ gives

$$(\text{FP}) = \frac{R_{\text{FP,in}} [K_3 (\text{FDP}) + 1](k_2 + k_{-1})}{[k_2 E^\circ - R_{\text{FP,in}}] k_1 K_3 (\text{ATP})(\text{FDP})}, \quad \text{for } (\dot{\text{FP}}) = 0 \quad (A1.8)$$

From (A1.8), we note first that no steady state is possible unless $k_2 E^\circ > R_{\text{FP,in}}$. This is reasonable, because $R_{\text{FP,in}}$ is the rate of supply and $k_2 E^\circ$ is the maximum attainable rate of removal. Unless $k_3 E^\circ > R_{\text{FP,in}}$, (FP) can only continue to accumulate. Further examination of (A1.8) shows that as (FDP)→0, the steady-state value of (FP) becomes infinite (without FDP to activate enzyme, FP continues to accumulate). As (FDP) gets very large, the steady-state value of (FP) approaches $R_{\text{FP,in}}(k_2 + k_{-1})/(k_2 E^\circ - R_{\text{FP,in}}) k_1 (\text{ATP})$. The curve is shown in Figure A1.5.

Solving (A1.5) for $(\dot{\text{FDP}}) = 0$ gives

$$(\text{FDP}) = \frac{k_1 k_2 K_3 E^\circ (\text{ATP})(\text{FP}) - k_4 (k_2 + k_{-1})}{k_4 K_3 (k_2 + k_{-1}) + k_1 k_4 K_3 (\text{ATP})(\text{FP})}; \quad \text{for } (\dot{\text{FDP}}) = 0 \quad (A1.9)$$

Since (FDP) cannot be negative, $(\dot{\text{FDP}})$ cannot be zero unless

$$(\text{FP}) \geq \frac{k_4 (k_2 + k_{-1})}{k_1 k_2 K_3 E^\circ (\text{ATP})}$$

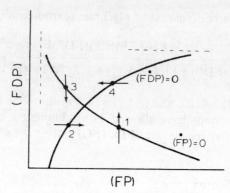

Figure A1.5. State-space diagram showing relation between FDP and FP.

As (FP) becomes larger, the value of (FDP) from (A1.9) asymptotically approaches $k_2 E°/k_4$. The shape of the locus for $(F\dot{D}P)=0$ is shown in Figure A1.5.

Equations (A1.8) and (A1.9) are the relations for the system that must simultaneously be satisfied at the steady state, which is at the crossing point of the two curves. We shall not, however, go through the algebra of solving the equations explicitly.

A1.7. Non-Steady-State Behavior

Directions in the state space may be deduced by the procedure in Section 7.7. First look at points 1 and 3. At these points we know that $(F\dot{P})=0$, and we need to determine $(F\dot{D}P)$. For point 1, (FP) is above the line for $(F\dot{D}P)=0$. This means that (X) is above the value needed to make $k_3(X) = k_4(FDP)$, so that $(F\dot{D}P) > 0$. Similar reasoning gives $(F\dot{D}P) < 0$ at point 3.

At points 2 and 4, $(F\dot{D}P) = 0$. Since at point 2 (FDP) is below the value required for $(F\dot{P}) = 0$, we may conclude that the enzyme is less activated than required, and that FP is accumulating, that is, that $(F\dot{P}) > 0$. Analogous reasoning gives $(F\dot{P}) < 0$ at point 4.

The nature of the feedback loop in Figure A1.4 and the directions shown in Figure A1.5 are compatible with the possibility of oscillatory behavior. Therefore, the model may tentatively be accepted as part of an explanation of observed oscillations in the concentrations of substances involved in glycolysis.

References

Atkinson, D. E. (1970). "Enzymes as Control Elements in Metabolic Regulation," in P. D. Boyer (Ed.), *The Enzymes*, 3rd ed., vol. I, Academic, New York.

Garfinkel, D., L. Garfinkel, M. Pring, S. B. Green, and B. Chance (1970). "Computer Applications to Biochemical Kinetics," *Ann. Rev. Biochem.* **39**: 473–498.

Higgins, J., R. Frenkel, E. Hulme, A. Lucas, and G. Rangazas (1973)."The Control Theoretic Approach to the Analysis of Glycolytic Oscillators," in B. Chance, E. K. Pye, A. K. Ghosh, and B. Hess (Eds.), *Biological and Biochemical Oscillators*, Academic, New York.

Koshland, D. E. (1970). "The Molecular Basis for Enzyme Regulation," in P. D. Boyer (Ed.), *The Enzymes*, 3rd ed., vol. I, Academic, New York.

Rapoport, T. A., R. H. Heinrich, and S. M. Rapoport (1976). "The Regulatory Principles in Erythrocytes *in vivo* and *in vitro*," *Biochem. J.* **154**: 449–469.

Stadtman, E. R. (1970). "Mechanisms of Enzyme Regulation in Metabolism," in P. D. Boyer (Ed.), *The Enzymes*, 3rd ed., vol. I, Academic, New York.

A2. CONTINUOUS CULTURE OF MICROORGANISMS

A2.1. General Background and Problem Formulation

Continuous culture procedures are used in the laboratory as a tool in basic research and in industry for the large-scale "harvesting" of microorganisms or the manufacture of metabolic products. Figure A2.1 is a sketch of a very simple continuous culture system.

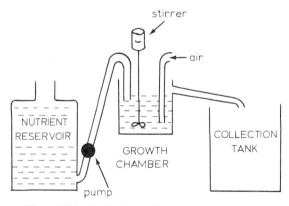

Figure A2.1. A simple continuous culture system.

The operation of the system involves feeding nutrients at a controlled rate. Note that the volume of material in the growth chamber remains constant, so that the rate at which nutrient is pumped is the same as the rate at which effluent is collected. Concentration of organisms in the effluent depends upon the rate of pumping and the rate of growth. In its turn, the rate of growth of the microorganism population depends on the size of the population at any instant and on the concentration of nutrient. The steady state of the microorganism population can be controlled by limiting the nutrient concentration. When this type of control is used, the system is called a *chemostat*.

For the sake of simplicity, we assume that we have a chemostat in which all nutrient materials except one are kept in plentiful supply. That is, the rate of growth of the population is limited by the concentration of a single nutrient substance, S.

The problem to be addressed is how to regulate the feed of nutrient solution so as to obtain maximum yield of microorganisms.

A2.2. System Components, Variables, and Their Pattern of Relationships

In defining the components of the system, it is convenient to divide the growth chamber into two components: nutrient and microorganism. The resulting component diagram is shown in Figure A2.2. Variables for the system are shown in Table A2.1. This choice of components and variables leads to the signal–flow graph of Figure A2.3.

Note that S_r, F, and V_g are "managerial" or "control" variables. They are not influenced by the other variables except through the managerial decision process. In what follows, we take V_g to be constant and try to determine how values for S_r and F might be chosen so as to get maximum yield of organisms per unit time, when the system is running at steady state.

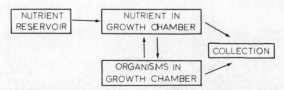

Figure A2.2. Component diagram for continuous culture system. Since this is not intended as a compartmental diagram, the arrows are not necessarily to be interpreted in terms of flow.

A2. CONTINUOUS CULTURE OF MICROORGANISMS

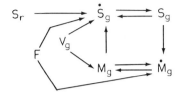

Figure A2.3. Signal–flow graph for chemostat system.

Table A2.1 Variables for Continuous Culture System

Variable	Symbol	Dimension
Nutrient concentration in reservoir	S_r	Mass/Volume
Nutrient concentration in growth chamber	$S_g(t)$	Mass/Volume
Amount of microorganism in growth chamber	$M_g(t)$	Number of organisms
Volume of growth chamber	V_g	Volume
Flow rate	F	Volume/time
Yield of microorganisms	$Y(t)$	Number/time
Rate of consumption of nutrient per organism[a]	μ	Mass/(number of organisms × time)

[a]This variable is introduced as a result of assumption a in subsection A2.4

A2.3. The Model Equations

For this system, it is most convenient to obtain a set of differential equations from the signal–flow graph, rather than to develop an input–output diagram. These equations necessarily involve assumptions that depend on the organisms and the nutrient, and whose validity may be a source of disagreement between workers.[3] We use three key assumptions:

Assumption a. A given amount of growth of the population of organisms requires a fixed amount of the limiting nutrient.

The effect of this assumption is that the rate of growth is proportional to the rate at which limiting nutrient is consumed. The assumption leads to the introduction of a new variable μ, as well as the parameters μ_{max} and κ (Table A2.2). The resulting equation is

$$\text{rate of growth} = \kappa \mu M_g \quad (A2.1)$$

[3]Indeed, the purpose of the model might be to test the validity of the assumptions.

Table A2.2 Parameters For Continuous Culture System

Parameter	Symbol	Dimension
Maximum rate of nutrient consumption per organism	μ_{max}	Mass/(number of organisms × time)
Amount of growth per nutrient consumed	κ	Number of organisms/mass
Nutrient concentration needed for half-maximal growth	K	Mass/volume

Assumption b. For very low values of S_g, the organisms consume nutrients as fast as possible, regardless of how much is available. For very large values of S_g, the rate at which organisms can use nutrient becomes saturated.

In other words, it is assumed that a plot of μ versus S_g has the appearance of Figure 5.7, so that

$$\mu = \frac{\mu_{max} S_g}{K + S_g} \qquad (A2.2)$$

where K is the value of S_g needed for half-maximal growth.

Assumption c. The rate of mixing within the growth chamber is fast enough so that concentrations of organisms and nutrient can always be considered uniform throughout the chamber.

An equation for $\dot{M}_g$ is formed from the relation,

$$\dot{M}_g = \text{rate of growth} - \text{rate of removal}$$

The rate of removal (amount removed per unit time) is the volume removed per unit time multiplied by the concentration,

$$\text{rate of removal} = F \frac{M_g}{V_g}$$

A2. CONTINUOUS CULTURE OF MICROORGANISMS

Combining this with (A2.1) and (A2.2),

$$\dot{M}_g = \kappa\mu M_g - F\frac{M_g}{V_g} \tag{A2.3}$$

$$= \left(\frac{\kappa\mu_{max}S_g}{K+S_g} - \frac{F}{V_g}\right) M_g \tag{A2.4}$$

The equation for $\dot{S}_g$ is

$$\dot{S}_g = \frac{1}{V_g} \times (\text{change in amount per unit time})$$

$$= \frac{1}{V_g} \times (\text{rate of addition} - \text{rate of removal} - \text{rate of use})$$

$$= \frac{1}{V_g}(FS_r - FS_g - \mu M_g) \tag{A2.5}$$

Finally, to obtain an equation for Y, we note that the concentration in the effluent is the same as the concentration in the growth chamber, M_g/V_g. The yield (amount per unit time) is, therefore,

$$Y = \frac{FM_g}{V_g} \tag{A2.6}$$

Equations (A2.4), (A2.5), and (A2.6) comprise the mathematical model for the system. Figure A2.4 shows the symbol–arrow graph.

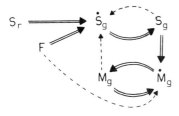

Figure A2.4. Symbol–arrow graph for continuous culture system.

A2.4. Model Behavior

In searching for states of the system in which $\dot{M}_g$ and $\dot{S}_g$ are zero, we exclude the uninteresting cases for which either F or M_g are zero, since this would mean zero rate of collection.

From (A2.4), we find that

$$S_g = \frac{FK}{\kappa\mu_{max}V_g - F}, \quad \text{for } \dot{M}_g = 0 \quad (A2.7)$$

Apparently, no nonzero steady state is possible if $F > \kappa\mu_{max}V_g$, since S_g cannot be negative. This might be interpreted as saying that if the flow is too fast, organisms are swept out so fast, the population can't possibly maintain itself.

From equations (A2.5) and (A2.2), we get

$$M_g = F\frac{S_r K + (S_r - K)S_g - S_g^2}{\mu_{max}S_g}, \quad \text{for } \dot{S}_g = 0 \quad (A2.8)$$

This expression cannot be evaluated at $S_g = 0$. However, for very low S_g, the numerator becomes dominated by the first term. Thus if S_g is to be very small, the value of M_g required to keep it constant gets larger and larger. This is reasonable, since the inflow of new nutrients is "trying" to increase S_g. For $S_g = S_r$, no M_g is required to keep S_g constant. The curve of M_g versus S_g for $\dot{S}_g = 0$ has a negative slope, whose magnitude decreases as S_g increases (see Figure A2.5).

In drawing Figure A2.5, it was assumed that $S_r > FK/(\kappa\mu_{max}V_g - F)$. Otherwise, equation (A2.7) tells us that the two curves do not cross; no steady state is possible for which $\dot{M}_g$ is nonzero. Biologically, this means

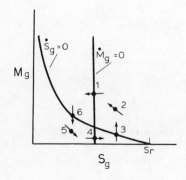

Figure A2.5. State-space diagram for continuous culture system.

A2. CONTINUOUS CULTURE OF MICROORGANISMS

that the organisms could never get enough nutrient to replace the cells being washed out.

By the methods of Chapter 7, one may find that the homeostatic index of the feedback loop is zero; the loop has no stabilizing effect on the position of the steady state.

The position of the steady state is determined by the values of κ, μ_{max}, V_g, S_r, and F. Assuming that S_r and F are chosen to allow a nonzero steady state, the directions shown in Figure A2.5 can be obtained using the procedure of Section 7.7.

Note that both the symbol–arrow and the state-space picture suggest the possibility of oscillations.

A2.5. Conditions for Maximum Yield

Finally, we should like to know how to choose S_r and F so that steady-state operation gives maximum yield per unit time. The first step is to solve (A2.7) and (A2.8) to get M_g^{ss} in terms of F and S_r,

$$M_g^{ss} = \kappa V_g \frac{\kappa \mu_{max} V_g S_r - F(K + S_r)}{(\kappa \mu_{max} V_g - F)}$$

This expression, together with (A2.6), gives

$$Y = \kappa F \frac{\kappa \mu_{max} V_g S_r - F(K + S_r)}{\kappa \mu_{max} V_g - F}$$

The dependence of Y on S_r is given by

$$\frac{\partial Y}{\partial S_r} = \kappa F \qquad (A2.9)$$

This is always positive, so the yield can always (according to this model) be increased by increasing S_r.

The derivative of Y with respect to F is a more complicated expression,

$$\frac{\partial Y}{\partial F} = \kappa \frac{(\kappa \mu_{max} V_g - F)[\kappa \mu_{max} V_g S_r - 2F(K + S_r)] + F[\kappa \mu_{max} V_g S_r - F(K + S_r)]}{(\kappa \mu_{max} V_g - F)^2}$$

The value for F that gives maximum yield can be found by setting this expression to zero (See subsection 8.3a), which means setting the numerator to zero. If the numerator is expanded and rearranged, it is seen to be a quadratic expression in F, with the two solutions,

$$F = \kappa\mu_{max}V_g\left(1 \pm \left(\frac{K}{K+S_r}\right)^{1/2}\right)$$

Since we already know that F must be less than $\kappa\mu_{max}V_g$, the proper solution is

$$F^* = \kappa\mu_{max}V_g\left(1 - \left(\frac{K}{K+S_r}\right)^{1/2}\right) \qquad (A2.10)$$

We find, therefore, that maximum yield can be obtained by making S_r as high as is consistent with solubility and engineering consideration and then setting F according to equation (A2.10). Note, however, that if S_r is high enough, it will no longer be rate limiting, so assumption a will no longer be applicable.

References

Maleck, I., and Z. Fencl (Eds.) (1966). *Theoretical and Methodological Basis for Continuous Culture of Microorganisms*, trans. J. Liebster, Academic, New York.

Novick, A., and L. Szilard (1950). "Experiments with the Chemostat on Spontaneous Mutations of Bacteria," *Proc. Natl. Acad. Sci.* **36**: 708–719.

Tempest, D. W. (1970), "The Continuous Culture of Microorganisms: 1. Theory of the Chemostat," in J. R. Norris and D. W. Ribbons (Eds.), *Methods in Microbiology*, vol. 2, Academic, New York.

Thingstad, T. T. (1974). "Dynamics of Chemostat Culture: The Effect of a Delay in Cell Response," *J. Theoret. Biol.* **48**: 149–159.

Williams, F. M. (1971). "Dynamics of Microbial Populations," in B. C. Patten (Ed.), *Systems Analysis and Simulation in Ecology*, Academic, New York.

Young, T. B., D. F. Bruley and H. R. Bungay, III (1970). "A Dynamic Model of the Chemostat," *Biotechnology and Bioengineering* **12**: 747–769.

A3. TEMPERATURE CONTROL

Temperature control in a homeothermic organism is based on a set of interconnecting control mechanisms. We begin by examining the operation of a single control mechanism in an inanimate thermostatic system and then extending the discussion to the case of two control mechanisms (one for heating and one for cooling) in the same system. This system serves as

A3. TEMPERATURE CONTROL

a crude physical model, whose examination suggests how to approach the study of temperature control in a physiological system.

A3.1. Description and Components

The inanimate system to be considered is a constant temperature water bath (Figure A3.1). In the drawing of Figure A3.1, it has been assumed that the bath is always run above environmental temperature, since a heating element has been provided, but no cooling element. Figure A3.2 is a component diagram showing where the input to each component comes from.

A3.2. The Variables and Their Pattern of Relationships

Variables for the system are shown in Table A3.1. Parameters are shown in Table A3.2.

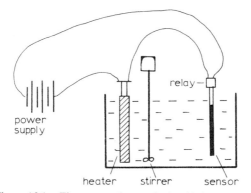

Figure A3.1. Thermostated water bath with heater only.

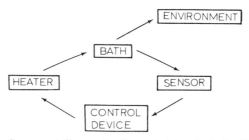

Figure A3.2. Component diagram for thermostated water bath with heater only.

Table A3.1 Variables for Thermostated Water Bath (Heating System Only)

Variable	Symbol	Dimension
Bath temperature (= sensor temperature)	$T_B(t)$	Temperature
Environment temperature	T_E	Temperature
Rate of heat loss from bath	$L_H(t)$	Energy/time
Rate of heat input to bath	$I_H(t)$	Energy/time
State of controller	$\delta(t)$	Dimensionless

Table A3.2 Parameters for Thermostated Water Bath (Heating System Only)

Parameter	Symbol	Dimension
High seting	T_H	Temperature
Low setting	T_L	Temperature
Maximum heater output	H	Energy/time
Surface area of bath	A	Area
Heat capacity of bath fluid	C	Energy/(Mass × temperature)
Thermal conductivity of the walls	Q	Energy/(area × temperature × time)
Mass of bath fluid	M	Mass

The last variable δ requires some special discussion. The cheapest controller we can buy would be a relay that would turn the heat completely on or completely off. That is, if we want the bath to hover around 60°C, the controller would turn the heat on (set δ to 1) when T_B gets, say, to 58°C and turn it off (set δ to 0) when T_B gets, say, to 62°C. If H is the capacity of the heater, then the heat it supplies at any time t is $H\delta$, where δ is the *state* (0 or 1) of the controller at time t.

It's easy to see that the temperature of such a system is going to oscillate even if the environmental temperature stays constant.

Another type of controller that might be used is a *proportional* controller. For such a controller, δ would vary smoothly from zero at T_H up to one at T_L, where it would be putting out maximum power. That is,

$$\delta = \frac{T_H - T_B}{T_H - T_L} \quad (A3.1)$$

Input from the heater would still be δH.

A3. TEMPERATURE CONTROL

The interrelations between the variables are diagrammed in Figure A3.3.

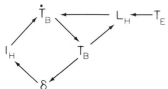

Figure A3.3. Signal–flow graph for thermostated water bath with heater only.

A3.3. The Model Equations and System Behavior

In order to construct a mathematical description of this system, we need to know something about the way in which heat transfer and temperature change depend on each other (see, for example, Casey, 1962, or Holwill and Sylvester, 1973). In this system, we're assuming that the heat transfer is limited by the conducting ability of the wall. The rate of heat loss per unit area is approximately proportional to the temperature difference (see Table A3.2 for definitions of Q and A),

$$L_H = QA(T_B - T_E)$$

The amount of temperature change produced by a given heat loss or gain depends on the substance being cooled or heated. Each substance has the capacity to absorb a different amount of heat energy for each degree of temperature change. The amount of heat energy per unit mass needed to change the temperature of the substance by one degree is called the heat capacity C. Since the total heat energy must be shared over all the mass of the bath fluid, we get

$$\dot{T}_B = \frac{1}{MC}(I_H - L_H)$$

$$= \frac{\delta H}{MC} - \frac{QA(T_B - T_E)}{MC} \quad (A3.2)$$

The details of the system behavior depend upon the behavior of δ, which depends upon the type of controller being used.

A3.3a. On–Off Controller. First, suppose $T_B > T_E$ and $\delta = 0$, so the heater is off. Equation (A3.2) becomes

$$\dot{T}_B = \frac{QA}{MC} T_E - \frac{QA}{MC} T_B \qquad (A3.3)$$

Since $T_B > T_E$, this is negative. It is most negative when $T_B = T_H$ (the high limit) and least negative when $T_B = T_L$ (the low limit). Therefore, the curve has the qualitative appearance of curve a in Figure A3.4.

When $T_B = T_L$, δ switches from zero to one, and we have

$$\dot{T}_B = \frac{1}{MC}[H - QA(T_B - T_E)] \qquad (A3.4)$$

From this equation, we can see that if T_B is to be kept from going below T_L, the heater must be large enough so that

$$H > QA(T_B - T_E)$$

If the heater is strong enough so that $\dot{T}_B$ remains positive even when $T_B = T_H$, then the picture is qualitatively like curve b in Figure A3.4. T_B rises to T_H; δ switches to zero, and the whole thing starts over.

In this case, we see that there is no steady state for the system in the narrow sense. However, if the heater is not strong enough so that $\dot{T}_B$ remains positive when $T_B = T_H$, then there is some value T^* between T_L and T_H, with the property that

$$H = QA(T^* - T_E)$$

In this case, T_B gradually increases, approaching T^* asymptotically. The temperature T^* would be a steady state for the system.

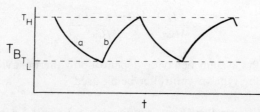

Figure A3.4. Temperature of bath using on–off relay, with $H > QA(T_H - T_E)$.

A3. TEMPERATURE CONTROL

A3.3b. Proportional Controller.
Combining equation (A3.1) with (A3.2) gives

$$\dot{T}_B = \frac{H}{MC} \frac{T_H - T_B}{T_H - T_L} - \frac{QA}{MC}(T_B - T_E) \qquad (A3.5)$$

Note that since $(H/MC)(T_H - T_L)$ is a constant, the heat input is proportional to the difference $(T_H - T_B)$.

Using equation (A3.5), we find that when $T_B = T_L$, $\dot{T}_B$ is given by equation (A3.4) as before (providing the heater is strong enough); when $T_B = T_H$, $\dot{T}_B$ is given by equation (A3.3), also as before. The difference is that $\dot{T}_B$ changes from negative to positive smoothly as T_B is raised, so that there must be a T^* at which $\dot{T}_B$ is zero.

To put it another way, equation (A3.5) is a difference between two positive terms. The first gets lower as T_B goes from T_L to T_H, whereas the second term gets higher. For $T_B < T^*$, we have $\dot{T}_B > 0$, so T_B is increasing. For $T_B > T^*$, we get $\dot{T}_B < 0$, so T_B is decreasing. Thus wherever T_B starts, the proportional controller causes the system to approach the steady-state value T^* asymptotically. Once "there," it puts out just enough heat to balance the heat loss.

A3.4. System with Heating and Cooling

If the water bath is to be all-purpose, it needs the addition of a cooling coil and a new controller element. As before, the controller can be an on–off type or proportional type. The description needs two more variables, and three new parameters, as shown in Table A3.3. The superscript$^{(c)}$ designates quantities specifically pertaining to the cooling system. We might want to define another variable, rate of heat *gain* from environment. However, we can just stick with L_H, rate of heat loss, and say that negative values of L_H mean heat *gain*. The symbol–arrow graph for the new system is shown in Figure A3.5.

Table A3.3 Variables and Parameters Needed for Cooling System

Quantity	Symbol	Dimension
Rate of heat removal by coil	$L^{(c)}(t)$	Energy/time
State of cooling coil controller	$\delta^{(c)}(t)$	Dimensionless
High set point for cooler	$T_H^{(c)}$	Temperature
Low set point for cooler	$T_L^{(c)}$	Temperature
Maximum cooler capacity	$H^{(c)}$	Energy/time

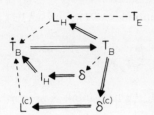

Figure A3.5. Symbol–arrow graph for thermostated water bath with heating and cooling systems.

Equation (A3.2) would be replaced by

$$\dot{T}_B = \frac{1}{MC}\left[\delta H - \delta^{(c)} H^{(c)}\right] - \frac{QA}{MC}(T_B - T_E) \qquad (A3.6)$$

where the second term is positive or negative depending on whether $T_B < T_E$ or $T_B > T_E$. Note that if $T_L^{(c)} < T_H$, we could have the heating system and the cooling system operating at the same time.

A3.5. Temperature Control in a Physiological System

A3.5a. General Background. The thermostated water bath system provides a conceptual framework within which relevant questions may be asked:

i. In the water bath, a uniform temperature is maintained by vigorous stirring. Clearly, the wall temperature is somewhere between T_B and T_E. In a physiological system, we must be concerned with the temperature *distribution* within the system and the relationships between the temperature of any two parts of the system. Furthermore, we want to consider the wall as *part* of the system, and we ask, Will the skin temperature always be between internal and environmental temperature?
ii. What are mechanisms for heat distribution within the system and are they temperature sensitive?
iii. Is there a single heating and a single cooling mechanism or are there several of each? What are they?
iv. If there are several of each, are they all activated by the same controller or is there a separate controller for activating each mechanism? In any case, what are the physiological counterparts of the controllers?
v. What are the mathematical characteristics of the controllers? Are they "on–off," "proportional," or something else?
vi. Is there a single temperature sensing center, several centers, or is the

A3. TEMPERATURE CONTROL

temperature over the whole body integrated to produce the heating and cooling reactions?
vii. Is there indeed anything that could be called a "temperature setting?"

Research on temperature control has provided partial, but not complete answers to these questions.

i. *Temperature distribution.* As a rule, skin temperature is between the internal body temperature and that of the environment, unless the body is being cooled by sweating. In that case, the skin temperature is lower than either.

ii. *Mechanism for heat distribution.* The main mechanism for heat distribution is blood flow. This mechanism is not only temperature sensitive, but it is also part of the controlling mechanism. Cold stress causes the blood vessels near the surface to constrict, thereby decreasing the conveyance of heat to the surface and increasing the effective thickness of the insulating layer. Heat stress causes the blood vessels near the surface to dilate, with the reverse effect. This, however, is effective only in conjunction with sweating, which lowers the skin temperature. In an environment hotter than body temperature, blood vessel dilation without sweating would only lead to increased heat *absorption*.

iii. *Mechanisms for heating and cooling.* The primary, but not the only, mechanism for heat production is increased metabolic activity, including shivering. At rest, the human body produces approximately 70 kcal of heat per hour. Under cold stress, it is capable of multiplying this figure as much as fourfold. Production of heat is supplemented by heat conservation through constriction of surface blood vessels.

The principle mechanism for cooling is sweat (or panting in animals without sweat glands) coupled with increased heat flow through vasodilation. The human body is capable of removing nearly 300 kcal of heat/hr—or about four times the normal metabolic rate of heat production.

iv. *Nature of the controllers.* Discussion of the controllers must clearly involve the central nervous system. Readers who are interested in this aspect of the problem should find the review by Hensel (1973) a convenient introduction to the literature.

v. *Characteristics of the controllers.* The mathematical characteristics of the controllers can be studied only very indirectly. The evidence is that the controllers have on–off characteristics and proportional characteristics, as well as other mathematical characteristics. The mathematical characteristics of the controllers must be one of the cornerstones of any model of temperature control.

vi. *Temperature-sensing centers*. The number of places that the temperature is sensed, and the relative importance of each, is very much a matter of experimental research. It appears that the temperature of the skin and the temperature of the hypothalmus are the two most important. The manner in which the two temperatures combine to produce the heating and cooling responses is not completely settled. Benzinger (1969) described the following general pattern: Heating responses depend primarily on peripheral cold-reception, but are inhibited by sufficiently high hypothalmic temperature; cooling responses depend primarily on cerebral heating but are inhibited by low enough peripheral temperature.

vii. *The set point*. The question of the reality of the "set-point" partly revolves around the way in which "set-point" is defined. The engineering literature generally regards a set-point as something that is external to and independent of the automatic control system. Such a definition doesn't take us very far in a biological system. Hensel (1973) defines the set-point as "that value of the controlled variable (in this case temperature) at which the control action is zero." Such a definition may have value. It allows us to ask questions about the reproducibility of the point from one individual to the next and from one experiment to the next. Do each of the controlling mechanisms have the same set-point? It appears that the controlling mechanisms do not all have the same set-point, that set-points in different individuals of the same species are very close to each other, and that the set-points within one individual are very reproducible but undergo rythmic variations and may be displaced in response to various stimuli (see reviews by Hensel, 1973; Benzinger, 1969).

Temperature regulation is apparently of very high priority in the operation of most homeothermic organisms. Under cold stress, they maintain high body temperature even in face of starvation, rather than conserve energy. Under heat stress, the body maintains a high rate of sweat secretion even in the face of dehydration, and excessive vasodilation may induce failure of the blood pressure regulatory system. There are exceptions, of course. Hybernating animals allow their internal temperatures to go to very low levels; camels allow their internal temperatures to vary by several degrees so as to conserve water.

A3.5b. System Components and Variables. Within the conceptual framework provided by very simple water bath thermostat systems, we have been able to pose a few relevant questions. The partial answers to these

A3. TEMPERATURE CONTROL

questions provided by experimental investigation, allow us to draw the component diagram of Figure A3.6. A partial list of variables is shown in Table A3.4. Positive values for $\dot{H}_M$, $\dot{H}_B$, and $\dot{H}_S$ mean heat gained, whereas positive values for $\dot{C}_W$ and $\dot{C}_E$ mean heat lost. The variables may be arranged in a symbol–arrow graph as in Figure A3.7. In this model, the interaction with the environment comes about through radiative, conductive, and convective heat transfer, which are proportional to body surface area, and through evaporation of water, which also depends on body surface area.

A3.5c. Development of the Model Equations. Discussion of this rather complex system is limited to a consideration of how one might approach development of the equations for the controlling responses, based on a qualitative consideration of the accumulated experimental evidence. References are provided for readers who would like more detail.

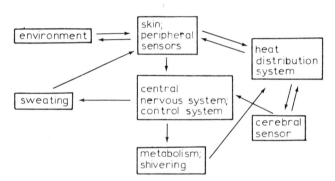

Figure A3.6. Component diagram for physiological temperature control system.

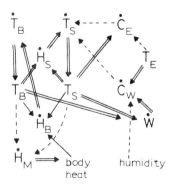

Figure A3.7. Symbol–arrow graph for a model of physiological temperature control.

Table A3.4 Variables for a Model of Physiological Temperature Control

Variable	Symbol	Dimension
Environmental temperature	$T_E(t)$	Temperature
Skin temperature	$T_S(t)$	Temperature
Cerebral temperature	$T_B(t)$	Temperature
Rate of metabolic heat production	$\dot{H}_M(t)$	Energy/time
Rate of sweat production	$\dot{W}(t)$	Mass/time
Rate of cooling by water evaporation	$\dot{C}_W(t)$	Energy/time
Rate of heat exchange between cerebral sensor and heat distribution system	$\dot{H}_B(t)$	Energy/time
Rate of heat exchange between surface and heat distribution system	$\dot{H}_S(t)$	Energy/time
Rate of heat loss from surface to environment (by conduction and radiation)	$\dot{C}_E(t)$	Energy/time

The model includes three kinds of control response: direct adjustment of rate of metabolic heat production (including shivering); sweating; vasodilation (reflected by $\dot{H}_S$). The indications seem to be that each of these is activated by a proportional controller, but that each responds both to a set point for skin temperature, T_S°, and one for cerebral temperature, T_B°. Thus a heating response would be activated only if $(T_B - T_B^\circ)$ and $(T_S - T_S^\circ)$ are *both* negative; a cooling response would be activated only if both are positive. Therefore, *any* response would require that the product $(T_B - T_B^\circ)(T_S - T_S^\circ)$ be positive. Now, if we label an arbitrary response by the letter R and let R_0 be the "normal" value for R (say, of metabolic heat production), the magnitude of response would be expressed by

$$R = R_0 + k(T_B - T_B^\circ)(T_S - T_S^\circ) \tag{A3.7}$$

where R would be a heating or cooling response depending upon the sign of the two factors. Recall that there are several possible heating and cooling responses. Each would have its own values for R_0, k, T_B°, and T_S°, allowing for the possibility that heating and cooling responses can be simultaneously activated.

It seems hardly necessary to state that the model represented by Figure

A3.7 does not account for all experimentally observed effects. The references below should serve as a good beginning for readers interested in exploring this topic further.

References

Benzinger, T. H. (1969). "Heat Regulation: Homeostasis of Central Temperature in Man," *Physiol. Rev.* **49**: 671–759.

Bligh, J., and R. E. Moore (Eds.)(1972). *Essays on Temperature Regulation*, Elsevier, New York.

Casey, E. J. (1962). *Biophysics, Concepts and Mechanisms*, Reinhold, New York.

Hammel, H. T. (1965). "Neurons and Temperature Regulation," in W. S. Yamamoto and J. R. Brobeck (Eds.), *Physiological Controls and Regulations*, Saunders, Philadelphia.

Hensel, H. (1973). "Neural Processes in Thermoregulation," *Physiol. Rev.*, **53**: 948–1017.

Holwill, M. E, and N. R. Sylvester (1973). *Introduction to Biological Physics*, Wiley, New York.

Riggs, D. S. (1970). *Control Theory and Physiological Feedback Mechanisms*, Williams & Wilkins, Baltimore.

Talbot, S. A., and U. Gessner (1973). *Systems Physiology*, Wiley, New York.

Whittow, G. C. (Ed.)(1970). *Comparative Physiology of Thermoregulation*, 3 vols., Academic, New York.

A4. REGULATION OF FOOD INTAKE AND CONTROL OF ENERGY STORES

The regulation of food intake in animals is an enormously complex control system, involving psychological, as well as physiological, components. LeMagnen (1971) points out that the regulation of food intake is a central and far-reaching problem in general biology. On the one hand, it is a physiological problem, concerning the constancy of the "milieu interieur," maintenance of the energy balance, thermoregulation; its control involves neurophysiology, sensory physiology, and physiological psychology. On the other hand, food intake is one of the major points at which species interact with one another. It is an important part of any serious attempt to model any ecological system, including human systems.

It is an extremely efficient control system. Bray and Campfield (1975) point out that in a single year, an average adult human male ingests over a million calories. If energy expenditure differed from energy intake by as little as 1%, there would be a weight change of about 3 pounds—well in excess of the yearly weight change for a normal adult male.

A4.1. General Background

In this appendix, we briefly review the important experimental results, and then examine a mathematical model that is based on a simplified view of the physiology. A list of references is provided for readers who wish to investigate this control system in more detail.

Starting even with the most general awareness of the digestive system, one might set down a diagram, such as that of Figure A4.1, as a framework within which to begin discussing the system. Figure A4.2 is based on the following experimental findings.

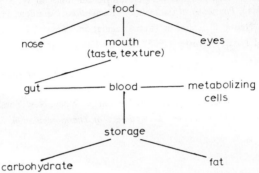

Figure A4.1. A preliminary diagram of a metabolic food processing system.

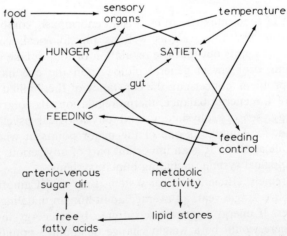

Figure A4.2. Diagram showing relationships between some factors involved in food intake control.

A4. REGULATION OF FOOD INTAKE

i. Feeding appears to be regulated by an "on–off" controller. The evidence seems to indicate that the brain is equipped with a "hunger center," which is responsible for turning on the feeding stimulus and a "satiety center," which is responsible for turning it off. Thus it is observed that most animals eat in discrete meals, the main exceptions being some marine filter-feeding animals.

ii. The most important direct physiological factor in activating the hunger center is thought to be the difference between arterial and venous glucose. A low difference induces feeding, although a high difference does not inhibit it. Thus the calorie content of a meal affects the length of time until the next meal but does not appear to influence its size. The effects of sustained energy use involve mobilization of lipid stores, which raises the plasma-free fatty acid content and lowers the arterio-venous glucose difference. The underlying control of food intake is thought to be a lipostatic mechanism, which strives to maintain constancy of bodily energy stores.

iii. A second set of factors that are important in activating the hunger center is the set of orosensory stimuli. The orosensory stimuli inform an animal about the presence of food, allow a distinction to be made between food items and nonfood items, and permit the establishment of an order of preferences among food items.

iv. Meal size is limited in part by filling of the gut, but this is not the whole story. It is also limited by the orosensory stimulus of eating and may be related to neural fatigue. Certain foods induce larger meal sizes and are said to be more "palatable." The orosensory stimulus that leads to a feeling of satiety appears to be food-specific. Thus overeating can be induced in many animals (including humans) by changing foods in the middle of the meal, presumably putting more reliance on the gut-filling mechanism for activating the satiety center.

v. Low temperatures increase feeding, whereas high temperatures decrease it. Most of this effect may be due to the tendency of homeotherms to increase or decrease metabolic heat production so as to maintain constant temperature. Aside from this general energy balance effect, however, there is thought to be a specific temperature effect that stimulates feeding at low temperatures and suppresses it under heat stress. It is supposed that this specific effect is related to the heat produced in the process of metabolizing the food and the high priority that homeothermic organisms place on temperature regulation. Whatever the reason, it is known that under heat stress the feeding drive in some animals is so completely suppressed that they suffer death by starvation.

The model we develop is based on the simplified conceptualization of Booth and Toates (1974). In this conceptualization, food intake is regulated by an on–off control, which acts to maintain the supply of readily available energy within certain bounds. Readily available energy is supplied both by metabolism of energy stores and by direct absorption from the gut. Thus in this model, gut-filling increases the supply of readily available energy and turns the feeding response off. Lowering of the energy stores and of the gut contents decreases the supply and stimulates the feeding response.

A4.2. Components and Variables

A component diagram for the model of Booth and Toates is shown in Figure A4.3. The pivotal variable in their formulation is the supply of readily available energy, which is expressed as the maximum rate at which the system can supply energy. The variables are shown in Table A4.1. The pattern of relationships is reasonably intuitive, and is shown as a symbol–arrow graph in Figure A4.4.

A4.3. The Model Equations

Since the controller is taken to be of the on–off type (see the discussion in

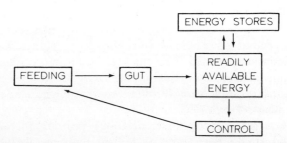

Figure A4.3. Component diagram for model of Booth and Toates (1974).

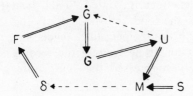

Figure A4.4. Symbol–arrow graph for model of Booth and Toates (1974).

A4. REGULATION OF FOOD INTAKE

Appendix A3), we need the three parameters δ, M_H, M_L, and F_M, shown in Table A4.2. At any given time, the rate of feeding is

$$F = \delta F_M \tag{A4.1}$$

If $\delta = 1$, the system is "on," and the organism feeds until $M = M_H$. At this point, δ switches to zero, and the feeding system is "off" until $M = M_L$.

According to this model, the available metabolizable energy is simply the sum of that available from energy stores and from the gut, so

$$M = U + S \tag{A4.2}$$

The net rate of gut filling at any time is simply the difference between the feeding and emptying rates. We get the differential equation,

$$\frac{dG}{dt} = F - U \tag{A4.3}$$

In this model, the rate of uptake depends upon gut content. The functional form of the dependence is based on an experimental correlative

Table A4.1 Variables for Food-Intake Control Model

Variable	Symbol	Dimension
Feeding rate	$F(t)$	Energy/time
Content of gut	$G(t)$	Energy
Rate of uptake from gut	$U(t)$	Energy/time
Rate of breakdown of body stores	$S(t)$	Energy/time
Readily metabolizable energy (maximum rate of supply)	$M(t)$	Energy/time
State of on–off controller (zero for "off," one for "on")	$\delta(t)$	Dimensionless

Table A4.2 Parameters for Food-Intake Control Model

Parameter	Symbol	Dimension
Controller high setting	M_H	Energy/time
Controller low setting	M_L	Energy/time
Maximum feeding rate	F_M	Energy/time
Absorption coefficient	K	$(\text{Energy})^{1/2}/\text{time}$

model; rate of absorption is found to be proportional to the square root of gut contents,

$$U = K G^{1/2} \tag{A4.4}$$

In the use of their model, Booth and Toates allow the parameter K to vary during the day.

Finally, the variable S, which couples the food-intake system to the rest of the organism, is treated as an input, as shown in Figure A4.4. Positive S indicates that energy stores are being broken down and supplying readily available energy. Negative S indicates a net build-up of energy stores; negative S thus represents an output for the system being considered.

By simple substitution, we can collapse equations (A4.1) through (A4.4) to the following two equations, which involve only the variables G, $\dot{G}$, and M (recall S is being treated as net input)

$$\dot{G} = \delta F_M - KG^{1/2} \tag{A4.5}$$

$$M = KG^{1/2} + S \tag{A4.6}$$

Note that equation (A4.5) is a differential equation, whose form depends upon the value of δ (zero or one).

A4.4. Behavior of the Model

To take a specific type of situation for illustration, we suppose S is negative (as might be for a growing animal in which body stores are being built) and assume it is constant. This requires that M is always large enough to supply S, so that $M_L > S$. Furthermore, we assume that when the animal eats, it eats rapidly, so that $F_M \gg KG^{1/2}$ and also, $F_M \gg S$.

Since S is being taken as constant (negative), equation (A4.6) can be solved for $G^{1/2} = (M-S)/K$, so that the high and low settings for M can be translated into high and low settings G_H and G_L.

Now, suppose we start at time $t=0$ with $M = M_H$ (the animal is satiated), $G = G_H$, and $\delta = 0$. Then

$$\dot{G} = -KG^{1/2}$$

Thus G is clearly decreasing with a decreasing slope. This differential equation may be solved to get G as a function of time,

$$G(t) = \tfrac{1}{4}\bigl(2(G_H)^{1/2} - Kt\bigr)^2 \tag{A4.7}$$

A4. REGULATION OF FOOD INTAKE

Combining this with equation (A4.6) gives

$$M(t) = M_H - \tfrac{1}{2}K^2 t \quad (A4.8)$$

These are shown as the curves on the left-hand side of Figure A4.5.

When M reaches M_L (say at time t_1), the switch throws δ to one. Using the assumption that $F_M \gg KG^{1/2}$, equation (A4.5) becomes $\dot{G} = F_M$. Equations (A4.7) and (A4.8) become

$$G(t) = G_L + F_M(t - t_1) \quad (A4.9)$$

$$M(t) = K\left[\left(\frac{M_L - S}{K}\right)^2 + F_M(t - t_1)\right]^{1/2} + S \quad (A4.10)$$

(check to see that A4.10 agrees that $M(t_1) = M_L$). The shapes of the curves are shown on the right-hand side of Figure A4.5.

Although this is a highly simplified model of food-intake control, Booth and Toates found that it was able to reasonably simulate the feeding behavior of rats, with parameter values: $F_M = 1000$ cal/minute (0.3 g of feed/min); $M_L = 18$ cal/min; $M_H = 60$ cal/min; daytime $K = 0.6$ cal$^{1/2}$/min; night time $K = 0.9$ cal$^{1/2}$/min. The difference in values of K is intended to account for the fact that rats are nocturnal animals.

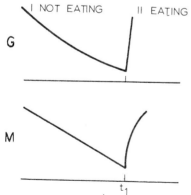

Figure A4.5. Qualitative character of time curves for gut energy G and available energy M.

References

Anand, B. K. (1961). "Nervous Regulation of Food Intake," *Physiol. Rev.* **41**: 677–708.

Booth, D. A., and F. M. Toates (1974). "A Physiological Control Theory of Food Intake in the Rat," *Bul. Psychonomic Soc.* **3**: 442–444.

Bray, G. A., and L. A. Campfield (1975). "Metabolic Factors in the Control of Energy Stores," *Metabolism, Clinical and Experimental* **24**: 99–117.

Brazier, M. A. B. (Ed.)(1962). *Brain and Behavior, Vol. 2, The Internal Environment and Alimentary Behavior*, Amer. Inst. of Biol. Sci., Washington, D. C.

Brobeck, J. R. (1960). "Food and Temperature," *Recent Progr. Hormone Res.* **16**: 439–466.

Code, C. F. (Ed.)(1967). "Alimentary Canal, Control of Food and Water Intake," *Handbook of Physiology*, Section 6, vol. I, American Physiological Society, Washington, D. C.

Conrad, H. R. (1965). "Symposium on Factors Influencing the Voluntary Intake of Herbage by Ruminants: Physiological and Physical Factors Limiting Feed Intake," *J. Anim. Sci.* **25**: 227–235.

Hamilton, C. L. (1965). "Control of Food Intake," in W. S. Yamamoto and J. R. Brobeck (Eds.), *Physiological Controls and Regulations*, Saunders, Philadelphia.

Lepkovsky, S. (1975). "Regulation of Food Intake," *Adv. Food Sci.* **21**: 1–69.

Koong, L. J., and H. L. Lucas (1973). *A Mathematical Model for the Joint Metabolism of Nitrogen and Energy*, N. C. State Univ., Inst. of Statistics Mimeo Series No. 882, Raleigh.

LeMagnen, J. (1971). "Advances in Studies on the Physiological Control and Regulation of Food Intake," *Progr. Physiolog. Psychol.* **4**: 203–259.

Tepperman, J., and J. R. Brobeck (Eds.) (1960). "Symposium on Energy Balance," *Amer. J. Clin. Nutrit.* **8**: 527.

A5. SIMPLE PREY–PREDATOR ECOSYSTEM

A5.1. General Background

In this section, we are concerned with a very simple prototype model, consisting of a single predator, which feeds on a single prey, which feeds on plant material. The system is an extension of that considered in Chapter 7 and is represented in Figure A5.1, which is clearly designed to demonstrate that artistic ability is not a prerequisite for these activities.

Figure A5.1. Representation of a simple prey–predator system.

A5. SIMPLE PREY–PREDATOR ECOSYSTEM

In Section 7.5, a simple Volterra model is considered for this system. It is found that the Volterra model has an obvious "defect"; namely that for zero predator population, the model predicts that prey will multiply without bound. In this section, we remedy this "defect" by introducing terms to account for intraspecies competition.

A5.2. The Variables and Their Pattern of Relationships

The system consists of three components, or subsystems: vegetation, prey, and predator (Figure A5.2). A list of variables is shown in Table A5.1. Note that number of prey and number of predators are taken to be different dimensions for the system. The pattern of relationships is represented by the symbol–arrow graph of Figure A5.3. The arrow from P to $\dot{P}$ has been left unsigned since, as we shall see, this depends upon the balance between the intrinsic rate of increase and the competition effect. The discussion is limited to the prey–predator interaction, so that the vegetation is taken as part of the environment.[1] Environmental effects are assumed to be constant.

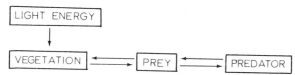

Figure A5.2. Component diagram for simple prey–predator system.

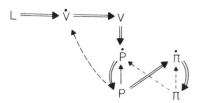

Figure A5.3. Symbol–arrow graph for simple prey–predator system.

Table A5.1 Variables for the Prey–Predator Ecosystem

Variable	Symbol	Dimension
Light intensity	$L(t)$	Energy/time
Plant material	$V(t)$	Mass
Prey	$P(t)$	Number of prey
Predators	$\Pi(t)$	Number of predators

[1]Simulation of the system, including the vegetation subsystem, is discussed in Section 9.3.

A5.3. The Model Equations

The starting point is taken to be the set of equations (7.18), whose rationalization is developed in Chapters 5 and 7. These equations are

$$\frac{dP}{dt} = \alpha_p - \beta_p P \Pi \tag{A5.1a}$$

$$\frac{d\Pi}{dt} = -\alpha_\pi \Pi + \beta_\pi P \Pi \tag{A5.1b}$$

Following the spirit of Section 6.7, a direct competition term in the prey equation might be proportional to P^2, to give equation (A5.2a).

For the predators, two alternative forms for a competition term might be rationalized. On the one hand, one might be inclined to say that predators are just naturally ornery; when two such creatures meet, they are likely to fight out of sheer cussedness. Such a premise leads to the introduction of a term proportional to Π^2. On the other hand, one might be inclined to assume that an event of direct competition takes place only if two predators happen to meet up with the same prey. This would suggest a term proportional to $P\Pi^2$. Using the second of these alternative premises, the equations for the system become

$$\frac{dP}{dt} = \alpha_p P - \beta_p P \Pi - \gamma_p P^2 \tag{A5.2a}$$

$$\frac{d\Pi}{dt} = -\alpha_\pi \Pi + \beta_\pi P \Pi - \gamma_\pi P \Pi^2 \tag{A5.2b}$$

The parameters and their dimensions are summarized in Table A5.2. The influence of the environment is not shown explicitly but is implicitly incorporated into the parameters.

A5.4. Steady States

As in Section 7.5, we have the uninteresting steady state, $(\Pi_0, P_0) = (0, 0)$. We next check the possibility of a steady state (Π_1, P_1), with $P_1 = 0$, but $\Pi_1 \neq 0$. From equation (A5.2b), we see that this requires $\alpha_\pi \Pi$ to be zero, which is implicitly assumed not to be the case. Thus the type of steady state being sought does not exist, which simply means that a population of predators cannot be sustained without a supply of prey.

However, we can find a steady state (Π_1, P_1) for which $P_1 \neq 0$, while

A5. SIMPLE PREY–PREDATOR ECOSYSTEM

Table A5.2 Parameters for the Prey–Predator Ecosystem

Parameter	Symbol	Dimension
Prey intrinsic rate of increase	α_p	Time^{-1}
Predator intrinsic rate of increase	α_π	Time^{-1}
Predation coefficient for prey	β_p	(Number of predators)$^{-1}$ time^{-1}
Predation coefficient for predators	β_π	(Number of prey)$^{-1}$ time^{-1}
Competition coefficient, prey	γ_p	(Number of prey)$^{-1}$ time^{-1}
Competition coefficient, predator	γ_π	(Number of prey)$^{-1}$ (number of predators)$^{-1}$ (time)$^{-1}$

$\Pi_1 = 0$. In this case, equation (A5.2a) can be solved for P_1;

$$P_1 = \frac{\alpha_p}{\gamma_p} \tag{A5.3}$$

Note that with $\Pi = 0$, this is simply the direct competition model of Section 6.6.

Next we look for a balance point (Π_2, P_2) with both populations nonzero. Setting the two derivatives equal to zero and assuming neither population size is zero, equations (A5.2) become

$$\alpha_p - \beta_p \Pi - \gamma_p P = 0 \quad \text{(if } \dot{P} = 0\text{)} \tag{A5.4a}$$

$$\alpha_\pi - \beta_\pi P + \gamma_\pi P\Pi = 0 \quad \text{(if } \dot{\Pi} = 0\text{)} \tag{A5.4b}$$

Solving equation (A5.4a) for P gives the condition under which dP/dt is zero for any *fixed* value of Π. Similarly, equation (A5.4b) gives the condition under which $d\Pi/dt = 0$ for any *fixed* P. We get

$$P = \frac{\alpha_p}{\gamma_p} - \frac{\beta_p}{\gamma_p} \Pi; \quad (\dot{P} = 0) \tag{A5.5a}$$

$$\Pi = \frac{\beta_\pi}{\gamma_\pi} - \frac{\alpha_\pi}{\gamma_\pi} \frac{1}{P}; \quad (\dot{\Pi} = 0) \tag{A5.5b}$$

A plot of equation (A5.5a) is simply a straight line with negative slope. One end is at $P = \alpha_p/\gamma_p$ for $\Pi = 0$, and the other is at $\Pi = \alpha_p/\beta_p$ for $P = 0$.

From equation (A5.5b), we find that as the "fixed" value of P gets large, predators become limited by their own competition, and Π approaches β_π/γ_π asymptotically. As P gets smaller, the prey population is able to support fewer and fewer predators, until $P = \alpha_\pi/\beta_\pi$, at which point no predator population can be supported. The plots of the two equations are shown in Figure A5.4.

The point of intersection of these two curves is the coexistence steady state (Π_2, P_2). Note that in drawing Figure A5.4, it was assumed that $\alpha_\pi/\beta_\pi < \alpha_p/\gamma_p$ and that $\alpha_\pi/\beta_\pi < \beta_\pi/\gamma_\pi$. Unless these conditions are met, the two curves do not intersect for positive values of P and Π; there is no coexistence steady state. The first of these inequalities simply says that the prey population must be inherently successful enough to support the predators. The second inequality states that competition between predators must not be so intense that they kill themselves off.

Accepting these assumptions, the simultaneous equations (A5.5) can be solved for Π_2 and P_2, although we shall not do so. The procedure is straightforward (but messy), except that it involves a quadratic expression only one of whose roots is acceptable.

An expression for the homeostatic index can be obtained directly from equations (A5.5)

$$\text{H.I.} = -\left(\frac{\partial P}{\partial \Pi}\right)\left(\frac{\partial \Pi}{\partial P}\right)$$

$$= \left(-\frac{\beta_p}{\gamma_p}\right)\left(+\frac{\alpha_\pi}{\gamma_\pi} \cdot \frac{1}{P^2}\right)$$

$$= +\frac{\beta_p \alpha_\pi}{\gamma_p \gamma_\pi} \cdot \frac{1}{P^2} \qquad (A5.6)$$

We have a negative feedback loop, with a positive homeostatic index. The index, however, is not a constant. This is a result of the fact that by equation (A5.5b), the dependence of Π on P is nonlinear; the slope of the curve (Figure A5.4) is, therefore, not constant. In such a case, the H.I. is given by the negative of the product of the slopes *at the steady state*. Examination of either Figure A5.4 or expression (A5.6) shows that the insulating capacity of the feedback loop, as measured by the homeostatic index, is less for steady states that involve larger amounts of prey.

A5. SIMPLE PREY–PREDATOR ECOSYSTEM

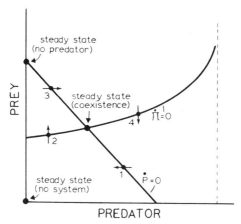

Figure A5.4. Locating the steady states and the directions in the state space for the simple prey–predator system described by equations (A5.2).

A5.5. Non-Steady-State Behavior

In order to arrive at the directions shown in Figure A5.4, we may reason on the basis of the curves of $\dot{P}=0$ and $\dot{\Pi}=0$. At points 1 and 3, we have $\dot{P}=0$, so the direction must be parallel to the Π-axis. Since it cannot be that $\dot{\Pi}=0$ (if it were, this would be a steady state), then Π must be either increasing or decreasing. Now, in the P dimension, point 1 is below the line for $\dot{\Pi}=0$. That is, P is less than the amount needed to maintain the predators at the indicated level. Therefore, the predator population must be decreasing; $\dot{\Pi}<0$. An analogous argument shows that $\dot{\Pi}>0$ at point 3.

At points 2 and 4, we have $\dot{\Pi}=0$; the direction must be parallel to the P-axis. At point 2, Π is less than that needed to limit the prey to the indicated level, and they must increase, so $\dot{P}>0$. Analogously, at point 4, the predators are in excess of the level required to just balance the prey population, so $\dot{P}<0$.

The state-space directions suggest the possibility of sustained oscillations. In order to see if the symbol–arrow graph confirms this, we need to know the sign of the $P\to\dot{P}$ arrow at the steady state. Increasing P increases the intrinsic rate of increase, leading to a positive feedback, but also increases the competition and predation effect, giving rise to a negative feedback. To be more precise, we differentiate equation (A5.2a),

$$\frac{\partial \dot{P}}{\partial P} = \alpha_p - \beta_p \Pi - 2\gamma_p P \tag{A5.7}$$

and try to evaluate the sign of this expression at the steady state. Substituting (A5.5a), which must hold at the steady state, into (A5.7) gives

$$\left(\frac{\partial \dot{P}}{\partial P}\right)_{ss} = -(\alpha_p - \beta_p \Pi) \tag{A5.8}$$

From Figure A5.4, we see that the steady-state value of Π must always be less than α_p/β_p, so the right-hand side of (A5.8) must be negative. Since all feedback loops in the system are negative, we now expect that the system spirals inward toward the steady state, and that the individual variables exhibit damped oscillations. Further mathematical details may be found in Pielou (1969).

It is especially important to notice that even a very slight alteration to the structure of the Lotka–Volterra equations of Chapter 7, in the form of any small but nonzero competition term, has switched the qualitative behavior of the model from one of sustained oscillations to one of approach to a steady state. When the qualitative behavior of a model is sensitive to slight modifications in structure, small errors in the specification of the functional forms can lead to large errors in predictions of system behavior (see also the discussion of Section 4.12). Such models are said to lack *structural stability*. Since one is rarely certain as to the absolute correctness of the structure of the model, models that lack structural stability (such as the Lotka–Volterra model of Chapter 7) may be hueristically useful, but must be regarded with suspicion when taken to represent actual real world systems.

A5.6. Comment on Models for Population Dynamics

In going from the model discussed in Section 7.5 to that discussed in this appendix, the complexity of the model has increased along with the capacity of the model to treat more complex prey–predator interactions. Nevertheless, a great many idealizations remain that limit the usefulness of the models. Some of the more glaring are:

a. Neglect of interactions involving other species.
b. Assumption that all individuals of a given species are stochastically identical. In particular, this leads to the neglect of the age structure of the population and the neglect of the delay between the time when an individual is added to the population and the time it becomes part of the reproducing population.

A5. SIMPLE PREY–PREDATOR ECOSYSTEM

c. Assumption of stochastic independence leads to neglect of sexual pairing, and the formation of cohorts, territoriality, and so on.
d. Environmental conditions, as represented by the parameters of the equations, are assumed to be constant.
e. The whole package of assumptions implied by the use of differential equations (subsection 5.1e).

It is up to the individual research worker to decide which idealizations are permissible for any given system. Some of the listed references describe models that deal with more complex interactions.

References

Holling, C. S. (1973). "Resilience and Stability of Ecological Systems," *Ann. Rev. Ecology and Systematics* **4**: 1–23.

Patten, B. C. (Ed.) (1971). *Systems Analysis and Simulation in Ecology*, 4 vols., Academic, New York.

Pielou, E. C. (1969). *An Introduction to Mathematical Ecology*, Wiley, New York.

Rescigno, A., and I. W. Richardson (1973). "The Deterministic Theory of Population Dynamics," in R. Rosen (Ed.), *Foundations of Mathematical Biology*, Academic, New York.

Watt, K. E. F. (1966). *Systems Analysis in Ecology*, Academic, New York.

APPENDIX B
MATHEMATICAL REVIEW

This appendix is a review of fundamental mathematical *concepts* (rather than manipulative skills) that are needed in the body of the book. This review is included because of the experience that unless a student's mathematical training *immediately* precedes an active interest in mathematical modeling, the mathematics becomes rusty through disuse. Further discussion on all of these topics may be found in the references of subsection 1.8b.

B1. Sets and Set Notation
B2. Relations Between Elements of Different Sets
B3. The Concept of Limits
B4. Continuous Functions
B5. Derivatives
B6. Integrals
B7. Logarithms

B1. SETS AND SET NOTATION

B1.1. The Notion of Sets

A *set* is simply a collection of things that anyone wishes to consider as a group. It may have a finite or an infinite number of members or *elements*. Usually, the reason for considering the elements as a group is that they have some property in common.

A set may be specified by either naming all the elements (only possible for finite sets) or by stating the property or properties that are membership

B1. SETS AND SET NOTATION

requirements. In either case, if the set is "well defined," the specification provides an absolute criterion for membership. That is, given any object, real or conceptual, we have a criterion by which we can say, "this *is* a member," or "this is *not* a member."

The usual notation uses curly braces to indicate that a set is being defined. In the following examples, the vertical line stands for "such that" (a colon, semicolon, or comma are other symbols often used for the phrase, "such that"),

$A = \{$Pentagon Building, Lincoln Memorial, Washington Monument$\}$

$B = \{$Numbers $x | x > 0\}$

$C = \{$People $x | x$ is a citizen of U.S.A.$\}$

$D = \{$Organisms $x | x$ is a mammal$\}$

The set A contains exactly three elements. The set B ("the set of numbers x such that x is greater than zero") is the set of positive numbers and is infinite in size. Set C is very large, but finite. Set D is just the taxonomic class *Mammalia*.[1]

Membership in a set is indicated by the symbol $\in$; lack of membership by $\notin$. For example, referring to the set B above, the statement $4 \in B$ would be read, "the number 4 is an element of (or, a member of) the set B." It would also be true that $-4 \notin B$.

Sets with single elements are a frequent source of confusion. A set is a distinct entity in its own right. The set $\{4\}$ is *not* the same as the number, 4. Since the set B contains only numbers, it is not true that $\{4\}$ is an element of B.

The need for biologists to sometimes deal with sets that are not well defined has been recognized, for example, through the coining of "polytypic concepts" (Beckner, 1959, p. 22) and "fuzzy sets" (Zadeh, 1965). A *polytypically defined set* would be specified by the statement of a large number of properties. Membership in the set is determined by possession of a large, but deliberately unspecified, number of these properties. The concept of *fuzzy sets* recognizes *degrees* of membership, ranging from 0 (*not* a member) up to 1 (*is* a member). The set "tall people" might be an example.

---

[1] See Buck and Hull (1966) for a discussion on the use of set notation to represent the Linnaean Hierarchy.

B1.2. Subsets

A subset is a *part* of a set. We indicate that one set is a subset (or part) of another set with the symbol $\subset$. *The symbol opens to the larger set.* For example, the following statements would be true,

$$\{4\} \subset \{\text{number } x \mid x > 0\}$$

$$\{\text{numbers } x \mid x > 4\} \subset \{\text{numbers } x \mid x > 0\}$$

A formal definition for the notion of subset would be:
$S_2 \subset S_1$ if every element of the set S_2 is also a member of the set S_1.

B1.3. Proper Subsets and Set Equality

If every element of S_2 is also an element of S_1 ($S_2 \subset S_1$), then one of the following must be true:

a. S_1 contains some elements that S_2 does not. In this case, S_2 is called a *proper subset* of S_1.
b. S_1 and S_2 contain exactly the same elements. In this case, the two sets are said to be *equal*.

Note that the *only* requirement for set equality is that both sets contain the same elements, with nothing said about the order or manner of specification. So,

$$\{1,4,9\} = \{9,4,1\}$$

Furthermore, double specification does not confer double membership, so

$$\{1,1,4,9\} = \{1,4,9\}$$

B1.4. Set Union

The union of two sets S_1 and S_2 is a new set that contains all the elements that are in either (or both) of the original sets. It is written

$$S_1 \cup S_2$$

If we look at the two sets,

$$G = Gymnospermae = \{\text{cone-bearing plants}\}$$

$$A = Angiospermae = \{\text{flowering plants}\}$$

B1. SETS AND SET NOTATION

their union is

$$S = G \cup A$$
$$= Spermatophyta$$

The union of several sets $S_1, S_2, \ldots, S_n$ is a new set that contains every element that is in *at least one* of the original sets. It may be denoted by

$$S_1 \cup S_2 \cup \cdots \cup S_n \quad \text{or} \quad \bigcup_{i=1}^{n} S_i$$

B1.5. Set Intersection

The intersection of two sets S_1 and S_2 is a new set that contains the elements that are in *both* of the original sets. It is written

$$S_1 \cap S_2$$

For example, taking the intersection of the set *Mammalia* with the set {flying animals}, we find the result to be the set *Chiroptera* = {bats}.

The intersection of several sets $S_1, S_2, \ldots, S_n$ is a new set that contains the elements that are in *every one* of the original sets. It may be denoted by

$$S_1 \cap S_2 \cap \cdots \cap S_n \quad \text{or} \quad \bigcap_{i=1}^{n} S_i$$

B1.6. The Null Set; Disjoint Sets

If the sets S_1 and S_2 have no elements in common, it would seem that there would be no such set as $S_1 \cap S_2$. In general, it would be a great hardship to have to know if S_1 and S_2 have elements in common before knowing if our rules allow the use of $S_1 \cap S_2$ in formal, strictly logical expressions. Partly for this reason, we invent a new kind of set—the set with no elements in it. It is like an empty box. It is given the symbol $\emptyset$, and is called the *empty set* or the *null set*. It has the properties that for any set S,

$$S \cup \emptyset = S \tag{B1.1}$$

$$S \cap \emptyset = \emptyset \tag{B1.2}$$

The first of these follows because $\emptyset$ brings nothing new. The second follows because the new set $(S \cap \emptyset)$ contains only elements which are both in S and in $\emptyset$. There are no such elements, so the new set is just $\emptyset$.

Two sets S_1 and S_2 are said to be *disjoint* if they have no elements in common. That is, if

$$S_1 \cap S_2 = \emptyset$$

B1.7. Universal Set; Set Compliment

If I write $\{x|x>0\}$, it is usually clear that of all the kinds of elements I could be discussing, I have limited the discussion to real numbers. On the other hand, if I write $\{x|x \text{ that fly}\}$, it may not be clear whether I am writing about animals, vehicles, pilots, or something else. Most frequently, the context makes it clear that the set that is being specified is not selected directly from the collection of all possible entities, but that a limited "universe of discourse" is agreed upon in advance. A moment's reflection should convince the reader that this is exactly what is normally done in ordinary conversation. When we say that something has been "quoted out of context," we mean that a statement that has been made in the framework of one "universe of discourse" has been incorrectly applied to another.

The universal set, then, is the large set from which we agree to select the specified sets. Sometimes the universal set is clear from the context, other times it must be explicitly specified.

Sometimes we wish to refer to a particular set only for the purpose of excluding it. We want a set that has all elements (of the universal set) *except* the one we specify. The *compliment* of any set S is the set of things (that are in the universal set) that are *not* in the set S. It may be written

$$S^c \text{ or } \overline{S}$$

Obviously, if $S = \{x|x \text{ that fly}\}$, the meaning of S^c is very different depending upon the context, that is, the universal set.

If we denote the universal set U, then for any set S it is true that

$$S \cup S^c = U \tag{B1.3}$$

$$(S^c)^c = S \tag{B1.4}$$

$$S \cap S^c = \emptyset \tag{B1.5}$$

B1.8. Partitioning the Universal Set

From equation (B1.3), we see that the pair of sets S and S^c divides U so that every element of U is in S or in S^c, but it cannot be in both. The set $\{S, S^c\}$, whose elements are the sets S and S^c, is called a *partition* of the universal set U.

The universal set can be partitioned into as many pieces as we have elements. Figure B1.1 shows U partitioned into the n sets $S_1, S_2, \ldots, S_n$. In order for the collection $\{S_1, S_2, \ldots, S_n\}$ to qualify as a partition of U, two properties are required:

a. Every element of U must be in one of the S_i; that is,

$$\bigcup_{i=1}^{n} S_i = U \tag{B1.6}$$

b. No element of U can be in more than one of the S_i; that is, every pair of sets S_i and S_j must be disjoint,

$$S_i \cap S_j = \emptyset, \text{ for } i \neq j \tag{B1.7}$$

Using this terminology, the collection $\{Gymnospermae, Angiospermae\}$ is a partition of the set *Spermatophyta*. The collection, $\{Urochordata, Cephelochordata, Vertabrata\}$ is a partition of the set *Chordata*.

References

Beckner, M. (1959). *The Biological Way of Thought*, University of California Press, Berkeley.

Buck, R. C., and D. L. Hull (1966). "The Logical Structure of the Linnaean Hierarchy," *Systematic Zoology* **15**: 97-111.

Zadeh, L. A. (1965). "Fuzzy Sets," *Information and Control* **8**: 338.

B2. RELATIONS BETWEEN ELEMENTS OF DIFFERENT SETS

In this discussion, the names of sets are designated by upper case letters; set elements are designated by lower case letters. The sets that we are

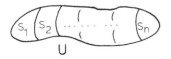

Figure B1.1. Partitioning the set U.

dealing with are sets of values of measured variables. For example, suppose we are interested in the relation between height and "diameter at breast height" of a certain kind of tree, say yellow pine. Why? That's beside the point, but we might need the relation as part of a model of the flow of energy or the balance of biomass in the forest. Or we might be working for a sawmill, and the relation is of monetary interest.

We let

U = set of all possible measurements of length

$H = \{h|h \text{ is a height}\}$

$W = \{w|w \text{ is a width}\}$

These sets contain *all possible* outcomes of the measuring process. If we let

$H_T = \{h_T | h_T \text{ is the height of some tree}\}$

$W_T = \{w_T | w_T \text{ is the width of some tree}\}$

then

$H_T \subset H$ and $W_T \subset W$

B2.1. Ordered Pairs; Ordered n-tuples

The outcome of a single observation might be, for example, the pair of numbers (10 ft, 4 in.). In this case, the units used make it rather clear that the first number is from H, whereas the second number is from W. Without the specification of the units, the order might not have been so obvious and would have had to be specified explicitly.

DEFINITION B2.1.　ORDERED PAIR

An ordered pair is a pair of elements in which the first element always comes from a specific set, and the second element always comes from a specific set.

The set of *all possible* ordered pairs with the first element from one set and the second from another, is called the *Cartesian product* between the two sets.

DEFINITION B2.2.　CARTESIAN PRODUCT

If A and B are two sets, then the Cartesian product of A and B, written $A \times B$, is

B2. RELATIONS BETWEEN ELEMENTS OF DIFFERENT SETS

the set

$$A \times B = \{(a,b) | a \in A \text{ and } b \in B\}$$

Note that $A \times B$ is *not* the same as $B \times A$. Referring to ordered pairs in the Cartesian product $H \times W$, it whould be clear that the pair $(10,4)$ is not intended to be the same as the pair $(4,10)$.

In order for two ordered pairs to be equal, they must be equal *element by element*.

DEFINITION B2.3. EQUALITY OF ORDERED PAIRS

Two ordered pairs (a_1, b_1) and (a_2, b_2) are considered equal if and only if $a_1 = a_2$ and $b_1 = b_2$.

There is no reason not to extend the concept of ordered pairs to ordered groups of elements from any arbitrary number of sets, say n of them. If we number the sets $S_1, S_2, \ldots, S_n$, and let "s_1" stand for some element of S_1, "s_2" stand for some element of S_2, and so on, then such an "ordered n-tuple" might be written, $(s_1, s_2, \ldots, s_n)$.

For example, if we look simultaneously at the outcome of three measurements, the ordered triple (3-tuple) might be convenient. In such a case we need to be concerned with the set of all possible ordered triples. Definition B2.2 can be extended to give

$$S_1 \times S_2 \times S_3 = \{(s_1, s_2, s_3) | s_1 \in S_1 \text{ and } s_2 \in S_2 \text{ and } s_3 \in S_3\}$$

If we extend this to n sets, the result is that the *Cartesian product* of sets $S_1, S_2, \ldots, S_n$ is the set of all n-tuples that can be formed with the restriction that the first element somes from S_1, the second from S_2, and so on,

$$S_1 \times S_2 \times \cdots \times S_n = \{(s_1, s_2, \ldots, s_n) | s_1 \in S_1, s_2 \in S_2, \ldots, s_n \in S_n\}$$

B2.2. Ordered n-Tuples and Points in Space

In the definitions of ordered pairs and ordered n-tuples, there is no restriction that all the sets be different. Suppose, for example, both A and B are sets of real numbers. Then $A \times B$ is the set of *ordered* pairs (a, b) where a is a number and b is a number. Examples might be the ordered pairs $(-5, 14)$, $(3, 2)$, $(2, 3)$, and so on. Note that $(3, 2) \neq (2, 3)$.

Such ordered pairs are often associated with points in space. Since this is a convenient association to use, let's look at it for a moment. Most of us are familiar with the representation of real numbers as points on a line. Such a representation is often called the "real number line." If we draw two copies of the real number line at right angles to each other,[2] then any point in two-dimensional space can be represented as an ordered pair of real numbers, as in Figure B2.1. Here one copy of the real number line has been labeled X, and the other has been labeled Y. If now we have an ordered pair (x,y), where x is a real number and y is a real number, we proceed out on the X-axis a distance represented by x, then proceed parallel to the Y-axis a distance represented by y. The point that is arrived at, p in the figure, is taken to represent the ordered pair. The numbers x and y are the *coordinates* of the point.

Since this point can also be associated with the line (or vector) that goes from the origin [which represents $(0,0)$] to the point p (line r in Figure B2.1), the ordered pair itself is sometimes called a *vector* or a 2-vector.

In the same way, a point in three-dimensional space can represent an ordered triple. Here we need three copies of the real line (Figure B2.2) which are usually labeled X, Y, and Z. The elements of the ordered triple are the coordinates of some point in the 3-space. For the same reason as before, the ordered triple may be called a 3-vector.

Although we cannot draw them, we can *conceptualize* spaces of any number of dimensions. For an ordered n-tuple, we need to conceptualize n copies of the real number line. The elements of the n-tuple are then thought of as coordinates of a point in "n-space." As with pairs and triples, the ordered n-tuple is often called an n-vector.

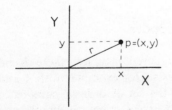

Figure B2.1. Real number line (a one-dimensional space) and the representation of an ordered pair of numbers as a point in two-dimensional space.

---

[2] The restriction of the two copies being perpendicular to each other is usually convenient, but not really necessary.

B2. RELATIONS BETWEEN ELEMENTS OF DIFFERENT SETS

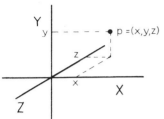

Figure B2.2. Representation of an ordered triple of numbers as a point in three-dimensional space.

B2.3. Relations and Functions

A single observation that gives a single ordered pair does not establish a relation between height and diameter. In order to establish such a relationship, we would need to make a number of observations. The purpose is to determine whether there is a pattern to the occurrence of the observed ordered pairs. If there were no relation between the height and diameter of the trees, then any ordered pair in the Cartesian product $H_T \times W_T$ could be observed as well as any other. The observations are likely to show, however, that only those ordered pairs occur in which both elements are large or both are small. That is, as a result of the relation between the two variables, the observations are likely to be restricted to a subset of the Cartesian product. Specification of the relation involves the specification of the subset of ordered pairs that can occur. Just as with the specification of any set, this can be done either by listing all the elements, which are the allowed ordered pairs, or by giving a property that all elements must have. In most (but not all) mathematical work, the property is stated as a mathematical equation. For example, examination of the observations might lead to the assertion,

$$h = 20w$$

The subset of allowed ordered pairs would then be

$$\{(h,w) \mid h \in H_T, w \in W_T, h = 20w\} \subset H_T \times W_T \tag{B2.1}$$

Since we would not expect this rule to hold absolutely, the assertion of the rule is equivalent to asserting that deviations are due to inaccuracy of measurement and accidental variability.

A mathematical equation then is a way of selecting an allowed subset of the Cartesian product of two sets. Two particular types of mathematical relationships are of special importance: functions and one-to-one functions.

B2.3a. Functions. A *function* is a rule that associates any member of some set (called the *domain* of the function) with a *unique* member of another set (called the *range* of the function).

If X and Y are two sets with elements x and y, we may write $y = f(x)$ (or $y = y(x)$) to indicate that there is a rule that associates every element of X with a unique element of Y. The rule is *named f* (or y). An example might be the equation,

$$y = x^2 \tag{B2.2}$$

Here, the domain set is the set of all real numbers, the range set is the set of nonnegative real numbers, and equation (B2.2) is the rule of association. By contrast, the equation,

$$y^2 = x^2$$

does *not* define a function, since for any value of x (except zero), there are two possible values for y; the rule of association does not specify a unique value for y.

Example. If y is the maximum concentration of oxygen that a sample of blood holds and x is the hemoglobin concentration, then $y = f(x)$.

Example. Suppose A and B are two chemical compounds that react according to the chemical equation,

$$2A + B \rightarrow \text{product}$$

If we let c_A and c_B be the concentrations and r be the rate of reaction, then

$$r = kc_A^2 c_B, \qquad k \text{ is a constant}$$

Here, we have *three* sets,

$$C_A = \{c_A | c_A \text{ is a concentration of } A\}$$
$$C_B = \{c_B | c_B \text{ is a concentration of } B\}$$
$$R = \{r | r \text{ is a reaction rate}\}$$

if we write

$$r = f(c_A, c_B)$$

B2. RELATIONS BETWEEN ELEMENTS OF DIFFERENT SETS

then the domain set for the function is the Cartesian product, $C_A \times C_B$. Each ordered pair of this set uniquely determines a rate r, which is an element of the range set R.

Full specification of the function requires three things:

a. Specification of the domain set.
b. Specification of the range set.
c. Specification of the rule by which the two sets are related.

The particular element of the domain set is often referred to as the *argument* of the function. The uniquely associated element of the range set is called the *value* of the function. Quite strictly, the function does not have a value until the argument is specified. In the statement $y = f(x)$, x is the argument (a particular element of the domain set), y is the value (a particular element of the range set), and f is the rule of association.

In the last example, the domain set is a Cartesian product, so that the argument is an ordered pair. It is not uncommon to have functions in which the domain set (or the range set or both) is a Cartesian product of a number of sets.

The way we have defined *function* does not require that either the domain set or the range set be numbers or n-tuples of numbers. If we are interested, for example, in setting a minimum time for heat sterilization of food, it would be quite correct to write

$$\text{time} = f(\text{food type, temperature})$$

Given a particular food type and sterilization temperature, there is a uniquely associated sterilization time required. Since food type is not a numerical variable, it would be difficult to express the rule of association in the form of an equation. Function values are, therefore, listed in tabular form, which is equivalent to the enumeration of the allowable ordered pairs, ((food type, temperature), time).

In the same vein, it might be useful in an appropriate context to write,

$$\text{type and severity of mental disorder} = f(\text{personality factors, specific biochemical factors, type and severity of stress experience, treatment received})$$

In this case not only are the variables nonnumeric, it is not even completely clear how to specify the elements of the domain or the range set.

What we have, in effect, is an attempt to list the factors involved (that is, identify the basic sets involved) as a prelude to more explicit specification.

As a further remark, it should be pointed out that it is not always either possible or desirable to specify every relevant variable of the argument (i.e., to specify every factor set of the Cartesian product, which is the domain set). It is generally implicitly understood that the stated relation is asserted to hold providing the levels of unspecified variables are held constant. In the case of the chemical reaction, such variables might be temperature, pressure, and concentration of other substances. In the case of food sterilization, such a variable might be the bacterial load in a given shipment of the food (that is, its state of cleanliness).

B2.3b. One-to-One Functions. If X is the set whose elements are x, and Y is the set whose elements are y, then to say that

$$y = f(x)$$

is to say that there is a set of ordered pairs, say

$$F = \{(x,y) \in X \times Y \mid y = f(x)\}$$

with the following property: Specification of the first element automatically identifies the second, just as specification of a child automatically identifies the second element of the pair (child, mother). However, since the same mother may correspond to more than one child, specification of the second element (mother) does not uniquely specify the first. In a patriarchal polygamous society, specification of the first element of the pair (wife, husband) automatically identifies the second, but the converse is not true. If the converse *is* true—that is, if specification of *either* element automatically identifies the other, then the relation between the two sets is a *one-to-one* function. In this case, the rule that goes from X to Y can simply be run backwards to go from Y to X. The backwards rule is called the *inverse function*. If the X to Y rule is named f, then the inverse function (backwards rule) is called f-inverse, written, f^{-1}.

Example a. If the function f is the *identity*, that is, $y = f(x) = x$, then the inverse, f^{-1} is the identity, $x = f^{-1}(y) = y$.

Example b. If $y = 1/x$, then the rule of association is to take the reciprocal of x, $f(x) = 1/x$. (Here the domain set is the set of nonzero real numbers.) The inverse function is $x = f^{-1}(y) = 1/y$.

B3. THE CONCEPT OF LIMITS

Example c. If $y = kx$ (k is a constant not zero), we have $f(x) = kx$. The inverse function would be $x = f^{-1}(y) = y/k$.

As illustrated by these examples, a one-to-one function is often expressed in the form of an equation that can be solved equivalently for x or y.

B3. THE CONCEPT OF LIMITS

A mathematical *limit* is something we get closer and closer to (as close as we like), without necessarily getting there. For example, consider the function,

$$y(x) = \frac{x^2 + x}{x} \tag{B3.1}$$

If we start with positive x and decrease x toward zero, this expression gets closer and closer to the value one. There is no value of x for which $y(x)$ is actually one; for $x = 0$, the expression is undefined, since division by zero is undefined. However, we can get it as close as we like to the value one by taking x close enough to zero. We say that $y(x)$ approaches the limit of one as x approaches zero. It would be written

$$\lim_{x \to 0} y(x) = 1 \quad \text{or} \quad y(x) \to 1 \text{ as } x \to 0$$

More generally, we would be interested in what happens to $y(x)$ as x gets close to some arbitrarily chosen value, which we'll call x_0. For example, it would be true that

$$\lim_{x \to x_0} \frac{(x - x_0)^2 + (x - x_0)}{(x - x_0)} = 1 \tag{B3.2}$$

If we set x_0 to zero, this expression becomes the same as that in (B3.1). This discusson can be summarized by the following general definition.

DEFINITION B3.1. LIMIT OF A FUNCTION OF ONE VARIABLE

If L_y is the limit of $y(x)$ as $x \to x_0$, that is, if

$$\lim_{x \to x_0} y(x) = L_y$$

then for any positive number ε, there is another positive number δ with the property that $y(x)$ will be within ε of L_y provided x is within δ of x_0.

This definition simply says that we can get $y(x)$ as close as we like to the limit L_y, provided we take x close enough to x_0.

Several times in the text, we want to study the behavior of a function as its argument becomes very large (approaches infinity). The idea is the same as in definition B3.1, but the wording has to be slightly different, because we cannnot speak of x as being within ε of infinity. Instead, we would say that $y(x)$ can be made to be as close as we please to its limit provided x is large enough.

Formally, this becomes

DEFINITION B3.2.   LIMIT AS ARGUMENT BECOMES FINITE

If L_y is the limit of $y(x)$ as $x\to\infty$, that is, if

$$\lim_{x\to\infty} y(x) = L_y$$

then, for any positive number ε, there is a number X with the property that $y(x)$ will be within ε of L_y provided x is larger than X.

As an example, if $y(x) = 1/x$, then $y(x) \to 0$ as $x \to \infty$.

Nothing in either definition B3.1 or definition B3.2 requires x to vary smoothly. We could just as easily speak of the limit of a sequence of values y_n, where n is restricted to integer values. For example, if $y_n = (n-1)/n$, then $y_n \to 1$ as $n \to \infty$.

Not every mathematical expression has a limit, and if it has, it might be infinite. Examples of expressions that have infinite limits are $1/x$, as x approaches zero, and $\tan\varphi$ as φ approaches $\pi/2$. As an example of an expression that has *no* limit, consider $(-1)^n$, where n is an integer. As the value of n gets larger and larger, the expression flips between -1 when n is odd and $+1$ when n is even. It never settles toward a limit.

A very useful special situation arises, when we wish to indicate that a function gets closer and closer not to some particular value, but to another function. For example, as x gets larger and larger, the expression in (B3.1) gets closer and closer to the function,

$$y'(x) = x$$

Although there is no value of x for which $y(x) = y'(x)$, we can get the difference between the two to be as small as we please by taking x large enough. Of course, as x becomes infinite, both $y(x)$ and $y'(x)$ become infinite. Our interest, however, is in what happens to $y(x)$ "on its way" to becoming infinite, that is, when x is very large. The situation is summarized by writing,

$$\lim_{x \to \infty} |y(x) - y'(x)| = 0$$

For large x, we, therefore, have $y(x) \cong y'(x)$, and we can approximate $y(x)$ by $y'(x)$, if it is convenient to do so (see Appendix C for further discussion on approximation).

B4. CONTINUOUS FUNCTIONS

The central idea of a *continuous function* is that of a function whose curve is a continuous piece, that is, it can be drawn without having to lift the pencil from the paper. This idea has been captured (that is, modeled) by the following definition.

DEFINITION B4.1. CONTINUITY AT A POINT

A function $f(x)$ is continuous at the point x_0 if

$$\lim_{x \to x_0} f(x) = f(x_0) \tag{B4.1}$$

This means that as x gets closer and closer to x_0, the function value gets closer and closer to its value at x_0; we can get $f(x)$ as close to $f(x_0)$ as we like by taking x close enough to x_0. In other words, the function approaches its value at x_0 in a smooth, continuous fashion.

DEFINITION B4.2. CONTINUITY IN AN INTERVAL

A function is said to be continuous in some given interval, if it is continuous at every point in the interval.

Figure B4.1a shows the graph of the continuous function $y(x) = x^2$. Figure B4.1b shows $y(x) = \tan x$ in the interval from zero to π. It is *not* continuous at $x_0 = \pi/2$. Indeed the limit as x approaches $\pi/2$ depends on the direction from which $\pi/2$ is approached. Finally, Figure B4.1c shows the graph of $y(x) = $ largest-integer-smaller-than-or-equal-to x (written $y(x) = [x]$). This function can only change by integer amounts, and so can be considered continuous only over intervals for which it is constant.

In biological systems, we often wish to deal with functions that represent the sizes of populations (of molecules or of organisms, etc.), and whose range set is the set of nonnegative integers. Apparently, such functions cannot be considered to be continuous.

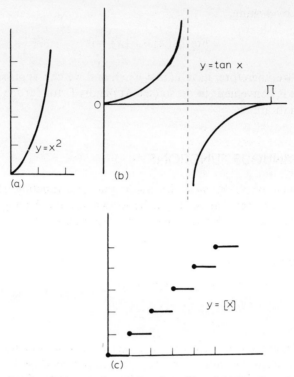

Figure B4.1. Graphs of (a) a continuous function; (b) and (c) functions that have discontinuities.

B5. DERIVATIVES

B5.1. Basic Concept

The *derivative* of a function at some particular point is just the slope of its curve at that point, that is, the slope of the tangent line at the point. The question is how to calculate it. If the function is a straight line, there is no problem. We merely select any two points (x_1, y_1) and (x_2, y_2), let $\Delta x = x_2 - x_1$ and $\Delta y = y_2 - y_1$. The slope at any point on the line is simply $\Delta y / \Delta x$. The problem is in getting the slope at some point on a curve whose slope is continually changing, say point A in Figure B5.1. The general procedure is to draw a straight line through the point A, that cuts the curve at another point, say A'. The coordinates of A are (x_0, y_0). The coordinates of A' are $(x_0 + \Delta x, y_0 + \Delta y)$. The slope of the line is simply $\Delta y / \Delta x$. Now, imagine

B5. DERIVATIVES

that the line is anchored so that it pivots on the point A. We swivel the line, so that the other point of contact moves closer to point A. This gives us a new value for Δy, for Δx, and for their ratio. We continue to swivel the line so that A' moves closer and closer to A. Finally, A' merges with A, and the line becomes the tangent line at point A. When this happens, we have $\Delta y = 0$ and $\Delta x = 0$, so we cannot compute the ratio $\Delta y / \Delta x$. But we *can* look at what happens to the ratio in the limit as the line approaches the tangent line, that is, as $\Delta x \to 0$. That limit is the slope, or *derivative*, of the curve at the point A.

To complete the formulation, we note that $y = y(x)$. So y_0 is the value of y when x is x_0, and $y_0 + \Delta y$ is the value of y when x is $x_0 + \Delta x$. A formal definition for the *derivative of y with respect to x* then becomes

$$\left(\frac{dy}{dx}\right)_{x=x_0} = \lim_{\Delta x \to 0^+} \frac{y(x_0 + \Delta x) - y(x_0)}{\Delta x} \qquad \text{(B5.1)}$$

where the notation $\Delta x \to 0^+$ indicates that Δx approaches zero from the positive side. The subscript on the left-hand side indicates that the slope of y versus x is to be evaluated at the point $x = x_0$. Other notations used to show that the derivative is to be evaluated at the particular point $x = x_0$ are $(dy/dx)_{x_0}$ and $dy(x_0)/dx$.

In getting the derivative at the point A, we made sure that the line always cut the curve at a point for which x was larger than its value at A. But it would not have made any difference if we chose to approach the tangent line using lines that cut the curve at points with x values smaller than at A. We would have arrived at the same tangent line and the same value for the derivative. Such is not the case if we look for the derivative at the point B. A moment's reflection should convince the reader that whereas it might make sense to discuss the slope immediately to the left or immediately to the right of point B, we cannot discuss "the slope" *at* point

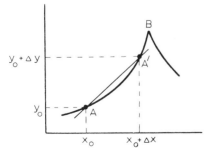

Figure B5.1. The derivative at the point A is obtained as the limit of the slope of the line from A to A', as A' approaches A.

B. Still less can we discuss the slope at the points of Figure B4.1 for which the curves are not continuous. That is, the *curve has to be continuous and sufficiently smooth in order for a derivative to exist.* Part of the formal work of differential calculus involves the closer definition of the phrase, "sufficiently smooth."

Note that the derivative dy/dx is *itself a function* of x. This property of the derivative is important in Section B5.2, as well as in Appendix B6.

When the independent variable represents time, the derivative is often denoted by a special notation,

$$\dot{y} \equiv \frac{dy}{dt}$$

A derivative with respect to time is often referred to as a rate of change or velocity. As an extension of this terminology, the derivative dy/dx may be referred to as "the rate of change of y with respect to x."

B5.2. Higher Derivatives

From Figure B5.1, we might obtain the derivative of each point (where it exists) and plot dy/dx versus x. The slope of the resulting curve would be the derivative of dy/dx with respect to x; that is, it would be the rate of change of the derivative. It is called the *second derivative* of y with respect to x and is generally denoted by d^2y/dx^2,

$$\frac{d^2y}{dx^2} \equiv \frac{d(dy/dx)}{dx} \tag{B5.2}$$

For a straight line, d^2y/dx^2 is zero, since the slope never changes. If the independent variable is time, then d^2y/dt^2 is the rate of change of the velocity and is called the *acceleration*.

In the same fashion, we could continue to third-order derivatives (rate of change of the second derivative) and to derivatives of still higher order.

B5.3. Partial Derivatives

When the argument of a function has several variables, we may look at the derivative of the function with respect to *one* of them. That is, we want to know how the function value changes as we change this one variable, while holding all the other variables *constant*. Consider, for example, a hillside, with x the value of the longitude and y the value of the latitude. We may

B5. DERIVATIVES

very well wish to discuss the slope of the hill in the x direction for a given value of y (partial derivative of height with respect to x), or the slope in the y direction for a given value of x (partial derivative of height with respect to y). We know that motion in any direction can be expressed as a combination of the motion in the x and the y directions. It follows (though we do not go into the details here) that the slope in any arbitrarily specified direction can be calculated once we know the *partial* derivatives in the x and the y directions.

In denoting partial derivatives, the "d" notation of equation (B5.1) is customarily replaced by the symbol ∂. In the calculation of partial derivatives, the same formulas may be used as for ordinary derivatives (see subsection B5.4), treating all variables as constant, except the indicated one. For example, for $f(x,y) = ax + by + cx^2 + dy^2 + exy$, we have

$$\frac{\partial f}{\partial x} = a + 2cx + ey \tag{B5.3}$$

$$\frac{\partial f}{\partial y} = b + 2dy + ex \tag{B5.4}$$

B5.4. Some Formulas for Evaluating Derivatives

Specific formulas for evaluating derivatives can be derived from the definition (B5.1). In the following formulas, c is taken to be a constant; n is taken to be an integer; e is the base of the natural logarithms:

i. $\dfrac{dc}{dx} = 0$

ii. $\dfrac{dx}{dx} = 1$

iii. $\dfrac{d(cx)}{dx} = c$

iv. $\dfrac{dx^n}{dx} = nx^{n-1}$ (except at $x = 0$ or $n = 0$)

v. $\dfrac{dcf(x)}{dx} = c\dfrac{df(x)}{dx}$

vi. $\dfrac{d[f(x) + g(x) + \cdots + h(x)]}{dx} = \dfrac{df(x)}{dx} + \dfrac{dg(x)}{dx} + \cdots + \dfrac{dh(x)}{dx}$

vii. $\dfrac{d\log_c f(x)}{dx} = \dfrac{\log_c e}{f(x)} \cdot \dfrac{df(x)}{dx}$

viii. $\dfrac{d\log_e f(x)}{dx} = \dfrac{1}{f(x)} \dfrac{df(x)}{dx}$

ix. $\dfrac{dc^{f(x)}}{dx} = c^{f(x)}(\log_e c)\dfrac{df(x)}{dx}$

x. $\dfrac{de^{f(x)}}{dx} = e^{f(x)}\dfrac{df(x)}{dx}$

B5.5. Derivative of a Function of a Function; The Chain Rule

Here, we consider the case of $y = y(x)$, where x is a function of another variable, say $x = x(z)$. We want dy/dz. Directly from (B5.1), we get

$$\frac{dy[x(z)]}{dz} = \lim_{\Delta z \to 0^+} \frac{y[x(z+\Delta z)] - y[x(z)]}{\Delta z}$$

Since $x(z+\Delta z) = x + \Delta x$ and since $\Delta z = (\Delta x) \cdot (\Delta z / \Delta x)$ and since $\Delta x \to 0$ as $\Delta z \to 0$ (assuming everything is adequately smooth), we get

$$\frac{dy}{dz} = \lim_{\Delta z \to 0} \frac{y(x+\Delta x) - y(x)}{\Delta x} \cdot \frac{\Delta x}{\Delta z}$$

or

$$\frac{dy}{dz} = \frac{dy}{dx} \cdot \frac{dx}{dz} \qquad (B5.5)$$

This can be extended, so that if a function f is a function of y, which is a function of x, which is a function of z, then

$$\frac{df(y(x(z)))}{dz} = \frac{df}{dy} \cdot \frac{dy}{dx} \cdot \frac{dx}{dz} \qquad (B5.6)$$

and so on, to any finite number of nested functions.

When we need to deal with functions or more than one variable, partial derivatives must be used. So, if we have a function $f(x,y)$, where $x = x(z)$

B6. INTEGRALS

and $y = y(z)$, we have (under appropriate conditions of smoothness)

$$\frac{df(x,y)}{dz} = \frac{\partial f}{\partial x} \cdot \frac{\partial x}{\partial z} + \frac{\partial f}{\partial y} \cdot \frac{\partial y}{\partial z} \tag{B5.7}$$

This also can be extended to any finite number of arguments

$$\frac{df(x_1, x_2, \ldots, x_n)}{dz} = \sum_{i=1}^{n} \frac{\partial f}{\partial x_i} \cdot \frac{\partial x_i}{\partial z} \tag{B5.8}$$

B6. INTEGRALS

B6.1. The Integral as a Sum

In Appendix B5, the derivative was introduced as the rate of change of one variable with respect to another. Derivatives are important when we are interested in the *rate* of accomplishment of some process. Integrals, on the other hand, measure the total accomplishment up to some given point. Suppose, for example, we wish to measure the total growth of a tree during some time interval Δt, and suppose for the moment that the rate of growth has a constant value r. The total growth during the interval is $r \cdot \Delta t$. If, however, the rate changes abruptly, then we must get $r \cdot \Delta t$ separately for each time interval and add the two, so that the total is $r_1 \cdot \Delta t_1 + r_2 \cdot \Delta t_2$. But if the rate changes in a continuous manner, this method of calculation is of no use. Nevertheless, we can get an approximate value for the total growth by breaking up the time into many intervals, say n of them, measuring the rate once in each interval and summing to get

$$G_n = \sum_{i=1}^{n} r_i \Delta t_i$$

The smaller the individual time intervals, the better the approximation will be (of course, the smaller the individual time intervals are, the more of them there will be). Note that this does not give an exact answer as long as the Δt_i are nonzero, because the rate is actually changing during each time interval. On the other hand, the expression becomes meaningless if the Δt_i are equal to zero. The way out of this problem is to use the limit as the Δt_i approach zero. Thus we are lead to define the integral of r with respect to t

starting from the initial time t_0, to some specific time, say t^*, as

$$G(t_0, t^*) = \int_{t_0}^{t^*} r \, dt = \lim_{\Delta t_i \to 0^+} \left(\sum_i r_i \Delta t_i \right) \quad (B6.1)$$

We see that the integral can be regarded as the "sum of an infinite number of infinitely small terms." The total height added to the tree in the interval between the time t_0 and the time t^* is given by the value of the integral. The values t_0 and t^* are called the limits of integration (an unfortunate double use of the word, "limit").

The general definition of an integral of a continuous function $f(x)$ in some designated interval exactly parallels the development of equation (B6.1). The interval is divided into subintervals Δx_i, and we denote by $f(x)_i$ the value of the function at any point within the ith interval. The integral of the function $f(x)$, as x varies over the interval from x_0 to x^* is defined as

$$\int_{x_0}^{x^*} f(x) \, dx = \lim_{\Delta x_i \to 0^+} \left(\sum_i f(x)_i \Delta x_i \right) \quad (B6.2)$$

B6.2. The Integral as an Area

Suppose we plot the function $f(x)$ versus x, as in Figure B6.1. The value $f(x)$ for any value of x is just the height of the curve at that vaue of x. The

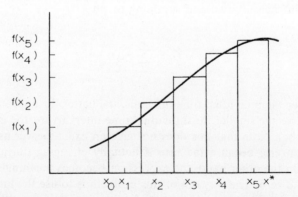

Figure B6.1. The area under the curve is approximated as the sum of the areas of the rectangles.

B6. INTEGRALS

value of $f(x)_i \cdot \Delta x_i$ is, then, just the area of the ith rectangle shown in the figure. We can get an approximation to the area under the curve from x_0 to x^* by adding up the areas of all the rectangles. The smaller and more numerous we make the rectangles, the better the approximation will be. We approach an exact value for the area in the limit as the width of each rectangle approaches zero, and the number of rectangles becomes infinite.

B6.3. Relation Between Derivatives and Integrals

In the example that began this section, the integral of r from t_0 to t^* was the height added to the tree during the time interval. On the other hand, at any time t, the value of $r(t)$ is just the derivative (rate of change) of the height. That is, if $h(t)$ is the height at time t,

$$r(t) = \frac{dh(t)}{dt}$$

Since the height at any specific time t^* would be the height at some starting time t_0 plus the height added from t_0 to t^*, we can write

$$h(t^*) = h(t_0) + \int_{t_0}^{t^*} \frac{dh}{dt} dt \tag{B6.3}$$

In general, if we have a continuous function $F(x)$, whose derivative (which is also a function of x) we designate by $f(x)$, the relation between them is expressed by the pair of equations

$$F(x^*) = F(x_0) + \int_{x_0}^{x^*} f(x) dx \tag{B6.4}$$

$$\left(\frac{dF(x)}{dx}\right)_{x=x^*} = f(x^*) \tag{B6.5}$$

In terms of Figure B6.1, $f(x)$ is the derivative of the area. If we write equation (B6.4) in the form

$$\int_{x_0}^{x^*} f(x) dx = F(x^*) - F(x_0) \tag{B6.6}$$

then we have a prescription for finding the integral of any given function $f(x)$. We determine the function F that has f for its derivative, evaluate F

at the limits of integration, and subtract. The function F is often called the *antiderivative*.

Determining the antiderivative is often very difficult and extensive tables of integrals have been published. (See any handbook of mathematical tables.) For the most part, such tables omit the integration limits, and are written simply in the form

$$\int f(x)\,dx = F(x)$$

B6.4. Some Rules for Evaluating Integrals

i. $\int_{x_0}^{x^*} cf(x)\,dx = c\int_{x_0}^{x^*} f(x)\,dx$

ii. $\int_{x_0}^{x^*} [f(x) + g(x)]\,dx = \int_{x_0}^{x^*} f(x)\,dx + \int_{x_0}^{x^*} g(x)\,dx$

iii. $\int_{x_0}^{x^*} f(x)\,dx = -\int_{x^*}^{x_0} f(x)\,dx$

iv. $\int_{x_0}^{x_1} f(x)\,d(x) + \int_{x_1}^{x_2} f(x)\,dx = \int_{x_0}^{x_2} f(x)\,dx$

B7. LOGARITHMS

B7.1. Definition and Basic Properties

A logarithm is just a power or exponent. That is, the meaning of the statement

$$2 = \log_{10} 100$$

is that 2 is the power to which 10 must be raised in order to get the number 100. In general, the statement

$$y = \log_b u \qquad (B7.1)$$

means that y is the power to which b must be raised in order to get the number u. It is equivalent to writing

$$b^y = u \qquad (B7.2)$$

B7. LOGARITHMS

That is,

$$b^{(\log_b u)} = u \tag{B7.3}$$

In the usual terminology, *y is the logarithm (or log) of u to the base b*.

In the defining relations (B7.2) and (B7.3), the base b can be any positive, real number. The value of the logarithm (y in equations (B7.1) and (B7.2)) can be any real number. The logarithm of a number u is thus defined only if u is a positive real number.

Several rules follow immediately from (B7.3). To get the log of the product $u \cdot v$, we use (B7.3) together with the simple properties of exponents,

$$u \cdot v = (b^{\log_b u})(b^{\log_b v})$$
$$= b^{\log_b u + \log_b v}$$

That is,

$$\log_b (u \cdot v) = \log_b u + \log_b v \tag{B7.4}$$

Similarly,

$$\frac{1}{u} = \left(\frac{1}{b}\right)^{\log_b u}$$
$$= b^{-\log_b u}$$

so

$$\log_b \frac{1}{u} = -\log_b u \tag{B7.5}$$

Other frequently used relations that are obtained from equation (B7.3) are

$$\log_b u^x = x \log_b u \tag{B7.6}$$

$$\log_b b = 1 \tag{B7.7}$$

$$\log_b 1 = 0 \tag{B7.8}$$

B7.2. Changing the Base

Suppose we know $\log_\alpha u$ but we are really interested in $\log_\beta u$. If we let $y = \log_\alpha u$ (known) and $x = \log_\beta u$ (to be found), then by definition, we have $u = \alpha^y$ and $u = \beta^x$, so

$$\alpha^y = \beta^x \tag{B7.9}$$

Now, if we take $\log_\alpha$ of both sides of (B7.9) and make use of (B7.6), we get

$$y \log_\alpha \alpha = x \log_\alpha \beta \tag{B7.10}$$

But it must always be true that $\log_\alpha \alpha = 1$. Therefore, (B7.10) leads to the general rule,

$$\log_\beta u = \frac{\log_\alpha u}{\log_\alpha \beta} \tag{B7.11}$$

Equation (B7.11) may be used with a table, say, of logs to base 10 to get logs to any other base. For example, to get $\log_2 15$, we note that $\log_{10} 15 = 1.1761$ and $\log_{10} 2 = 0.3010$, so $\log_2 15 = 1.1761/0.3010 = 3.9073$.

B7.3. Choosing the Base

The choice of base is purely a matter of convenience. The three most often used are base 10, base 2, and base e.

Base 10 logarithms (also called *common* logarithms) are the most convenient when logs are being used as an aid to base 10 arithmetic. Tables of base 10 logs are widely available.

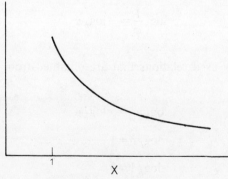

Figure B7.1. Graph of $y = 1/x$.

B7. LOGARITHMS

Base 2 logarithms are most often used in connection with information theory, communication theory, and computer arithmetic.

The base e is of special use in connection with calculus. The number e arises in the search for a function whose derivative is equal to itself; that is, with the property that

$$\frac{dy}{dx} = y \qquad (B7.12a)$$

The function that has property (B7.12a) is the function

$$y(x) = e^x \qquad (B7.12b)$$

where e is an irrational number, which cannot be represented exactly as a decimal. Its value is approximately 2.7182818284.

Another way of approaching the number e is through a search for the integral of the function $1/x$. If we draw the curve $f(x) = 1/x$ (Figure B7.1), then the area under the curve, starting with $x = 1$, is

$$\int_1^x \frac{1}{x} dx = \log_e x \qquad (B7.13)$$

The pair of equations (B7.12a) and (B7.12b) can be deduced from (B7.13) and vice versa. Taken together, they show the usefulness of the number e in integral and differential calculus.

Logarithms to base e are often indicated by the notation ln. They are often referred to as *natural* logarithms or as *Naperian* logarithms.

APPENDIX C
MATHEMATICAL APPROXIMATION

In Section 1.6, it is pointed out that mathematical approximation involves replacing one mathematical expression with another one, which is close to it under certain specified circumstances. For example, if

$$f(x) = x + x^2 \qquad (C.1)$$

then $f(x)$ can be approximated by the value of x if $|x|$ is small enough. This is true because

$$\lim_{x \to 0} |f(x) - x| = 0$$

It is often useful to introduce a new notation, which can concisely convey the idea that an approximation is being used as well as the circumstances governing the validity of the approximation.

DEFINITION C.1. LOWER ORDER OF MAGNITUDE

Suppose two functions $h(x)$ and $g(x)$ both have the real numbers for their domain and range sets, and that $g(x) \neq 0$ as long as $x \neq 0$. Then if the two functions are related in the following way

$$\lim_{x \to 0} \frac{|h(x)|}{|g(x)|} = 0 \qquad (C.2)$$

the relation is described by saying that $h(x)$ is of lower order of magnitude than $g(x)$. It is denoted by

$$h(x) = o[g(x)] \qquad (C.3)$$

Equation (C.3) is often read, "h is little oh of g."

MATHEMATICAL APPROXIMATION

Using this notation, we see that $x^2 = o(x)$. The approximation concerning expression (C.1) can, therefore, be written

$$f(x) = x + o(x) \tag{C.4}$$

Because of the way $o(x)$ is defined, expression (C.4) consicely implies the following: The first term on the right-hand side is to be used as an approximation to the function $f(x)$; the approximation is valid for small values of $|x|$; the approximation becomes increasingly precise as the magnitude of x decreases and becomes exact as x approaches zero.

The little oh notation can also be used to indicate comparisons between functions in limits other than as $x \to 0$, by explicitly specifying the limit intended. For example, expression (C.1) can be approximated by the term x^2 for x very large. The approximation becomes increasingly precise as x becomes larger and becomes exact as x becomes infinite. This can be expressed by writing

$$f(x) = x^2 + o(x^2) \quad \text{as} \quad x \to \infty$$

When the little oh notation is used without explicit specification of the limit intended, the limit is understood to be as $x \to 0$.

A property of the little oh relation that is especially important is that if we have a finite number of functions $h_1(x), h_2(x), \ldots, h_n(x)$, each one of which is $o[g(x)]$, then their sum is also $o[g(x)]$. That is, the sum of a large number of zeros is still zero; the sum of a finite number of negligible terms is negligible.

However, small values cannot be neglected when they are being multiplied or divided. For example, the values of xK or K/x are very sensitive to the value of x no matter what relation there is between the magnitudes of x and K. In Chapter 6, we encounter an expression of the type,

$$y = \frac{ax}{K + x}$$

The denominator is approximately K for $x \ll K$, or approximately x for $x \gg K$. The numerator, however, is always ax. As a result,

$$y \cong \frac{ax}{K}, \quad \text{small } x$$

$$y \cong \frac{ax}{x} = a, \quad \text{large } x$$

The remainder of this appendix discusses two widely used special types of approximation.

C.1. SERIES APPROXIMATIONS

The exponential function e^x can be written

$$e^x = 1 + x + \frac{x^2}{2!} + \frac{x^3}{3!} + \cdots + \frac{x^n}{n!} + \cdots \tag{C.5}$$

The right-hand side of this equation represents an infinite number of terms and is called an *infinite series*. However, if x is small, then $x^2/2!$ is smaller yet, $x^3/3!$ is still smaller; as n gets larger, $x^n/n!$ gets smaller. If x is small enough, terms past the first two can be neglected, and we can write

$$e^x = 1 + x + o(x) \tag{C.6}$$

so that the function is approximated by the few first terms of an infinite series.

The strategy is to express the function of interest as the sum of a series of an infinite number of terms that get smaller and smaller. The "rate" at which they get smaller depends on the value of x. If x is small enough, expression (C.6) would indicate a satisfactory approximation. For larger x, we would need to include more terms to get a satisfactory approximation. However, for *any* value of x, the function e^x can be approximated to any desired degree of closeness by taking a sufficiently large but *finite* number of terms of (C.5) even though an exact value would require an infinite number of terms. Examples of other common functions that can be approximated in this manner are

$$\sin x = x - \frac{x^3}{3!} + \frac{x^5}{5!} - \frac{x^7}{7!} + \cdots$$

$$\cos x = 1 - \frac{x^2}{2!} + \frac{x^4}{4!} - \frac{x^6}{6!} + \cdots$$

$$\ln(1+x) = x - \frac{1}{2}x^2 + \frac{1}{3}x^3 - \frac{1}{4}x^4 + \cdots$$

Many functions can be approximated by the leading terms of an infinite series; however, the question of when an infinite series can be approximated by its leading terms is often a difficult mathematical problem. In each case, it is necessary to establish that the sum of all neglected terms (there are always an infinite number of them) is negligible.

C.2. LINEAR APPROXIMATIONS

A frequently used type of approximation is the approximation of a smooth curve by a straight line, as illustrated in Figure C.1. The curve may be approximated in the vicinity of point A by the tangent line a, whose slope is $(df/dx)_A$. Note that as we get further away from point A, the line a becomes a worse approximation—the error (distance between the line and the curve) becomes larger. So, for a given permissible error, the sharpness of the curve limits the size of the neighborhood over which the line a is usable. The figure makes it clear that in the vicinity of point B, a different linear approximation would be used (line b).

Mathematical descriptions of straight lines are relatively easy to handle. Local linear approximations are, therefore, of great use in discussing changes in the value of a function that accompany small changes in the function argument. For example, using the equation of a straight line, we can get an approximate value of $f(x^*)$ for any x^* near x_A. If we know the value of $f(x_A)$ and the value of $(df/dx)_A$,

$$f(x^*) = f(x_A) + \left(\frac{df}{dx}\right)_A \cdot (x^* - x_A) + o(x^* - x_A) \tag{C.7}$$

Equation (C.7) indicates that the approximation becomes more precise as $|x^* - x_A|$ becomes smaller, finally becoming exact in the limit as $|x^* - x_A|$ approaches zero.

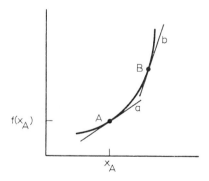

Figure C.1. Local approximation of a curve by straight lines.

APPENDIX D
LINEARITY AND THE USE OF AVERAGES

In the discussion of Section 3.6, it is suggested that an average shape factor might be used in equation (3.7). The questions are What kind of average should be used and How should it be determined?

By far, the most commonly used type of average is the arithmetic average or *arithmetic mean*. It is simply the sum of a whole bunch of numbers divided by the total number of numbers.

$$\bar{x}_{\text{arith.}} = \frac{1}{n} \sum_{i=1}^{n} x_i \tag{D.1}$$

Using this kind of average, the average surface area per animal would be

$$\bar{A}_{\text{arith.}} = \overline{\left(k_1 V^{2/3}\right)}_{\text{arith.}}$$

It would be nice if this could be written as

$$\bar{A} = \bar{k}_1 \bar{V}^{2/3}$$

That is, it is usually more convenient to be able to average each variable separately before combining them. The question is Does it make any difference whether we take averages before or after forming the mathematical expression; Is it true that $\overline{(kV^{2/3})} = \bar{k}\bar{V}^{2/3}$? Let's try it for two animals

$$\overline{kV^{2/3}} = \tfrac{1}{2}\left(k_1 V_1^{2/3} + k_2 V_2^{2/3}\right) \tag{D.2}$$

$$\bar{k}\bar{V}^{2/3} = \left[\tfrac{1}{2}(k_1 + k_2)\right]\left[\tfrac{1}{2}(V_1 + V_2)\right]^{2/3} \tag{D.3}$$

LINEARITY AND THE USE OF AVERAGES

If these are really the same mathematical expression in different disguise, then they have to be equal for every possible combination of values of k_1, k_2, V_1, and V_2. If we can find one combination for which (D.1) and (D.2) are *not* equal, then that will show that they are different mathematical expressions. So, let's suppose that $k_1 = k_2 = k$. Then we have

$$k \overline{(V^{2/3})} = \tfrac{1}{2} k (V_1^{2/3} + V_2^{2/3})$$

$$k (\overline{V})^{2/3} = k \left[\tfrac{1}{2}(V_1 + V_2) \right]^{2/3}$$

These are quite clearly not the same, so (D.1) and (D.2) are not the same.

A clue about what to try next comes from the logarithmic form of equation (3.7),

$$\log A = \log k + \frac{2}{3} \log V \tag{D.4}$$

This is the form of the equation that might be used to determine k from measured values of A and V; that is, if the animals are approximately geometrically similar, a plot of $\log A$ versus $\log V$ is a straight line of slope $2/3$ and intercept $\log k$. Taking the mean of $\log A$ gives

$$\overline{(\log A)}_{\text{arith.}} = \tfrac{1}{2}(\log A_1 + \log A_2)$$

$$= \tfrac{1}{2}\left(\log k_1 + \frac{2}{3}\log V_1 + \log k_2 + \frac{2}{3}\log V_2 \right)$$

$$= \tfrac{1}{2}(\log k_1 + \log k_2) + \frac{2}{3} \cdot \frac{1}{2}(\log V_1 + \log V_2)$$

$$= \overline{(\log k)}_{\text{arith.}} + \frac{2}{3} \overline{(\log V)}_{\text{arith.}}$$

In other words, if it is the logarithms whose averages are taken, then the average can be taken either before or after expression (D.4) is formed.

The next step is to see what happens if we take the antilog of these averages. Starting with A, we get

$$\overline{(\log A)}_{\text{arith.}} = \tfrac{1}{2}(\log A_1 + \log A_2)$$

$$= \log A_1^{\frac{1}{2}} + \log A_2^{\frac{1}{2}}$$

$$= \log(A_1 A_2)^{1/2}$$

So the antilog of $\overline{(\log A)}_{\text{arith.}}$ is $(A_1 A_2)^{1/2}$.

LINEARITY AND THE USE OF AVERAGES

In general, if we have a whole sequence of numbers $x_1, x_2, \ldots, x_n$ and if we take the antilog of the arithmetic mean of their logarithms, we get the nth root of their product:

$$\overline{(\log X)}_{\text{arith.}} = \frac{1}{n} \sum_{i=1}^{n} \log x_i$$

$$= \frac{1}{n} \log \prod_{i=1}^{n} x_i$$

$$= \log \left(\prod_{i=1}^{n} x_i \right)^{1/n}$$

Taking the antilog gives

$$\left(\prod_{i=1}^{n} x_i \right)^{1/n}$$

The nth root of the product of n numbers is called their *geometric mean*. It is the antilog of the arithmetic mean of their logarithms;

$$\bar{x}_{\text{geom.}} = \left(\prod_{i=1}^{n} x_i \right)^{1/n}$$

$$= \text{antilog} \, \overline{(\log X)}_{\text{arith.}}$$

Whereas it was *not* true that $\bar{A}_{\text{arith.}} = \bar{k}_{\text{arith.}} \bar{V}^{2/3}_{\text{arith.}}$, it *is* true that

$$\bar{A}_{\text{geom.}} = \bar{k}_{\text{geom.}} \bar{V}^{2/3}_{\text{geom.}} \qquad (D.5)$$

Now the question is, why could the logarithmic equation be written in terms of arithmetic means when the original one could not. The answer lies in the fact that the logarithmic equation, taken by itself, is the equation of a straight line. The connection is illustrated by Figure D.1. In this figure, there are plots of three different functions of X: $f_1(x)$, $f_2(x)$, and $f_3(x)$. Each has a different kind of curvature. The point marked with an asterisk represents the ordered pair $(\bar{x}_{\text{arith.}}, \bar{y}_{\text{arith.}})$. It lies on the straight line between

LINEARITY AND THE USE OF AVERAGES

(x_1,y_1) and (x_2,y_2). If the function relating x to y is a straight line function (in the figure, $f_2(x)$ is a straight line function), then the value of $f_2(\bar{x}_{\text{arith.}})$ is the same as $\overline{(f_2(x))}_{\text{arith.}} = \bar{y}_{\text{arith.}}$. If the function curves upward, as does $f_1(x)$, then the point $(\bar{x}_{\text{arith.}}, f_1(\bar{x}_{\text{arith.}}))$ lies below. That is $f_1(\bar{x}_{\text{arith.}}) < \overline{(f_1(x))}_{\text{arith.}}$. If the curvature is reversed, as for $f_3(x)$, then $f_3(\bar{x}_{\text{arith.}}) > \overline{(f_3(x))}_{\text{arith.}}$.

Now, the equation $A = kV^\alpha$ is a straight line only in special conditions ($\alpha = 1$ and k is constant). In such a special case, the stated relation holds between $\bar{A}_{\text{arith.}}$ and $\bar{V}_{\text{arith.}}$.

When a function involves only the products of variables, we can always get a straight line function by taking logarithms. From the arguments already presented, we should, therefore, expect that the *geometric mean* is useful.

To put it another way, any time we can convert a relation to a straight line relation, we expect that arithmetic means based on the new relation are useful. The geometric mean arises from a straight line function between logarithms of the variables.

Another frequently used mean is the *harmonic mean*, which is based on a straight line relation between reciprocals of variables. By way of example, let's look at the relation,

$$y = \frac{ax}{x+b} \tag{D.6}$$

where a and b are constants. Such an equation arises in describing enzyme action and other saturation phenomena. In Section 6.8, it is found that this equation has the kind of curvature shown by $f_3(x)$ in Figure D.1. We, therefore, expect that $y(\bar{x}_{\text{arith.}}) > \overline{(y(x))}_{\text{arith.}}$. Taking reciprocals of both sides, however, gives

$$\frac{1}{y} = \frac{x+b}{ax}$$

$$= \frac{x}{ax} + \frac{b}{ax}$$

$$= \frac{1}{a} + \frac{b}{a}\frac{1}{x}$$

If we take a and b as constants and make a number n of observations

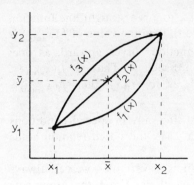

Figure D.1. Illustrating the relation between $f(\bar{x})$ and $\overline{f(x)}$. The value of $f(\bar{x})$ may be thought of as the measured value of y when x is $\bar{x}$; $\overline{f(x)}$ is the arithmetic mean of the two values $y_1 = f(x_1)$ and $y_2 = f(x_2)$.

(x_i, y_i), then $\overline{(1/x)}_{\text{arith.}}$ and $\overline{(1/y)}_{\text{arith.}}$ would be related by

$$\overline{\left(\frac{1}{y}\right)}_{\text{arith.}} = \frac{1}{n} \sum_{i=1}^{n} \frac{1}{y_i}$$

$$= \frac{1}{n} \sum \left(\frac{1}{a} + \frac{b}{a} \cdot \frac{1}{x_i}\right)$$

$$= \frac{1}{n} \sum \frac{1}{a} + \frac{1}{n} \sum \frac{b}{a} \cdot \frac{1}{x_i}$$

$$= \frac{1}{n} \cdot \frac{n}{a} + \frac{1}{n} \frac{b}{a} \cdot \sum_{i=1}^{n} \frac{1}{x_i}$$

$$= \frac{1}{a} + \frac{b}{a} \overline{\left(\frac{1}{x}\right)}_{\text{arith.}} \tag{D.7}$$

Now, it should be clear that the reciprocal of $\overline{(1/y)}_{\text{arith.}}$ is not $\bar{y}_{\text{arith.}}$. Indeed, it is given a special name, the harmonic mean.

DEFINITION. HARMONIC MEAN

The harmonic mean of a sequence of numbers $x_1, x_2, \ldots, x_n$ is

$$\bar{x}_{\text{harm.}} = \frac{n}{1/x_1 + 1/x_2 + \cdots 1/x_n} \tag{D.8}$$

provided it exists. It does not exist if any x_i is zero or if $\sum_{i=1}^{n} 1/x_i = 0$.

If we now take reciprocal of (D.7), we see that (D.6) expresses a relation

LINEARITY AND THE USE OF AVERAGES

between the harmonic means,

$$\bar{y}_{harm.} = \frac{a\bar{x}_{harm.}}{b + \bar{x}_{harm.}}$$

The discussion in connection with Figure D.1 is not meant to indicate that arithmetic means should only be used with straight line functions. For example, if a function reverses its curvature, randomly or regularly (such as in Figure D.2), then the arithmetic mean is a useful indication of the average behavior or "trend."

As an exercise, cite one pair of variables from your own field whose relationship would suggest the use of each of the kinds of averages: arithmetic mean, geometric mean, and harmonic mean.

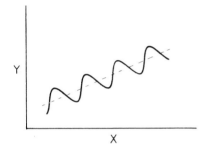

Figure D.2. A nonlinear function for which arithmetic averages might be useful.

APPENDIX E
DISCUSSION OF EXERCISES

1.2 The parameter a is always the value of y for $x=0$. Now, since $x \gg x^2$ for very small x, equation (1.3) is dominated by the $a + bx$ part for very small x. For very large x, it is dominated by the cx^2 term. Therefore, if you choose b to be positive, the curve starts out with positive slope as do both curves iii and iv. If c is positive, then the slope is increasingly positive as the cx^2 term takes over (curve iii). If $c < 0$, then the slope becomes negative as cx^2 takes over (curve iv). Try it with some actual numbers. Note that the qualitative appearance of the curve is being discussed in terms of the direction of the slope and the kind of curvature. Other qualitative characteristics that are often important involve existence of maxima, minima, and inflection points. In many cases of interest, the qualitative characteristics may be discussed in terms of the first and second derivatives.

1.5 It seems clear that of the three alternatives given, alternative b is the only one that has a chance of meeting assumption 2. Alternative a is least likely to meet either assumption. The author of the article, however, chose alternative a and thus generated what is, in my opinion, a "textbook example" of a mismatch between experimental conditions and mathematical assumptions.

3.1 a $60 \text{ miles/hr} = 60 \cdot \text{miles/hr} \cdot \dfrac{5280 \text{ ft/mile}}{3600 \text{ sec/hr}}$

$= 60 \times 1.4667$

$= 88 \text{ ft/sec}$

b $60 \text{ miles/hr} = 60 \text{ miles/hr} \cdot \dfrac{1.6 \text{ km/miles}}{\text{hr}}$

$= 60 \times 1.6$

$= 96 \text{ km/hr}$

DISCUSSION OF EXERCISES

$$\text{c} \quad 15 \text{ lb/in.}^2 = (15 \text{ lb/in.}^2)\left[\frac{0.45 \text{ kg/lb}}{(2.54 \text{ cm/in.})^2}\right]$$

$$= 15 \text{ lb/in.}^2 \frac{0.45 \text{ kg/lb}}{6.45 \text{ cm}^2/\text{in.}^2}$$

$$= 15 \cdot 0.0698$$

$$= 1.05 \text{ kg/cm}^2$$

$$\text{d} \quad 15 \text{ lb/in.}^2 = 15 \text{ lb/in.}^2 \frac{4.4 \times 10^5 \text{ dyn/lb}}{6.45 \text{ cm}^2/\text{in.}^2}$$

$$= 15 \times .6822 \times 10^5$$

$$= 1.02 \times 10^6 \text{ dyn/cm}^2$$

$$\text{e} \quad 32 \text{ ft/sec}^2 = 32 \text{ ft/sec}^2 \cdot \frac{30.48 \text{ cm/ft}}{\left[(1/60)(\min/\sec)\right]^2}$$

$$= 32 \text{ ft/sec}^2 \cdot \frac{30.48 \text{ cm/ft}}{(1/3600)\min^2/\sec^2}$$

$$= 32 \times 1.097 \times 10^5$$

$$= 3.51 \times 10^6 \text{ cm/min}^2$$

4.1 The "population" being sampled consists of the four *types*. It is sampling with replacement; neither 4 nor 4^2 combinations is enough to code for 20 amino acids. The minimum would be $4^3 = 64$.

4.2 Sampling without replacement gives $\binom{100}{10}$ possible different combinations of 10 from 100 flies. If each is equally likely, the probability of getting any *given* 10 is

$$1 / \binom{100}{10} = \frac{90!10!}{100!} = 5.8 \times 10^{-14}$$

4.3 If every rearrangement of the N molecules gave a different protein, we would have $N!$ different possibilities. Since for the jth type of amino acid, the n_j molecules may be permuted among themselves $n_j!$ ways without changing the type of protein, we get $N!/n_1!n_2!...n_{20}!$ different protein types. Another way of getting the same answer is to ask how many ways are there of dividing the "population" of N *positions* into 20 subpopulations, so that group 1 has n_1 elements, group 2 has n_2 elements, and so on, and then applying equation (4.12).

4.4 Since the F_1 all had round yellow seeds, then R must be dominant and Y

must be dominant. For F_2, we get

$$P(\text{round}) = \frac{3}{4}, \quad P(\text{wrinkled}) = \frac{1}{4}$$
$$P(\text{yellow}) = \frac{3}{4}, \quad P(\text{green}) = \frac{1}{4}$$

Assuming shape and color to be independent,

$$P(\text{round, yellow}) = P(\text{round})P(\text{yellow})$$
$$= \frac{3}{4} \cdot \frac{3}{4} = \frac{9}{16}$$

Similarly,

$$P(\text{round, green}) = \frac{3}{4} \cdot \frac{1}{4} = \frac{3}{16}$$
$$P(\text{wrinkled, yellow}) = \frac{1}{4} \cdot \frac{3}{4} = \frac{3}{16}$$
$$P(\text{wrinkled, green}) = \frac{1}{4} \cdot \frac{1}{4} = \frac{1}{16}$$

4.5 Since the probability of a mutation at a given site is 10^{-8}, the probability of *no mutation at a given site* is $(1-10^{-8})$. Since each site is independent, the probability of *no mutation at any site* is $(1-10^{-8})$ raised to the power 10^7. This may be estimated using logs,

$$P(\text{no mutation}) = (1-10^{-8})^{10^7}$$
$$\ln_e P(\text{no mutation}) = 10^7 \ln_e (1-10^{-8})$$

Using hint b, this becomes approximately

$$\ln_e P(\text{no mutation}) = -(10^7)10^{-8}$$
$$= -0.1$$

Taking the antilog gives

$$P(\text{no mutation}) = 0.9048$$

Since there must either be no mutation or at least one,

$$P(at\ least\ \text{one mutation}) = 1 - 0.9048$$
$$= 0.0952$$

DISCUSSION OF EXERCISES

5.1 In each case, it is necessary to find a time scale that is long relative to the rate at which elementary events take place. These might be

- **a** minutes for many chemical reactions.
- **b** hours, perhaps, but it would depend on the organism and the conditions.
- **c** months, perhaps, depending on conditions.
- **d** years.
- **e** billions of years.
- **f** days if compounded daily; years if compounded yearly.

6.4 J was *defined* as an amount per unit time per unit area, so it would have dimensions

$$(\text{amount})T^{-1}L^{-2}$$

Since C is amount per unit volume, $(\text{amount})L^{-3}$, we get, for equation (8.34),

$$(\text{amount})T^{-1}L^{-2} = -D \times \frac{(\text{amount})L^{-3}}{L}$$

D must, therefore, have dimensions of $L^2 T^{-1}$, that is, area per unit time.

8.1
- **a** $\varepsilon_z = -\frac{1}{y^2}\varepsilon_y$
- **b** $\varepsilon_z = x\varepsilon_y$
- **c** $\varepsilon_z = 2y\varepsilon_y$
- **d** $\varepsilon_z = (2y+1)x\varepsilon_y$
- **e** $\varepsilon_z = \frac{1}{y}\varepsilon_y$

8.3 For this exercise we need

$$\sum_i x_i = 50.9300 \qquad \sum_i y_i = 38.7200$$

$$\bar{x} = 5.093 \qquad \bar{y} = 3.872$$

$$\sum_i x_i y_i = 250.6400$$

$$\sum_i (x_i - \bar{x})^2 = 78.112 \qquad \sum_i (y_i - \bar{y})^2 = 42.0016$$

$$\sum_i (x_i - \bar{x})(y_i - \bar{y}) = 53.4400$$

These give

$$a = 0.3876$$
$$b = 0.6841$$
$$r = 0.9330$$

8.4 Taking $\varepsilon_i = 0.1y$, we get

$$\sum_{i=1}^{} \frac{1}{\varepsilon_i^2} = 208.9789$$

Thus $q = 10/\sum \varepsilon_i^{-2} = 0.0479$.
Letting $w_i = q/\varepsilon_i^2$ gives the following set of weights:

4.79, 1.87, 1.48, 0.76, 0.50, 0.18, 0.20, 0.10, 0.21, 0.10

Applying equation (8.28) gives

$$a = 0.310$$
$$b = 0.662$$

8.5 The lines obtained from exercises 8.2, 8.3, and 8.4 should all be plotted and compared. Note that exercise 8.4 gives more importance to accurately fitting the smaller values.

9.1 A position would have to be assigned that stores the constant 1. Then we might have

FETCH X
ADD 1
STORE X

9.2 Suppose $X = 4$ and we choose $Q = 1$, then

$$R_1 = \frac{4}{1} = 4$$

$$Q_2 = \frac{4+1}{2} = 2.5$$

$$R_2 = \frac{4}{2.5} = 1.60$$

$$Q_3 = \frac{2.50 + 1.60}{2} = 2.05$$

$$R_3 = \frac{4}{2.05} = 1.95$$

$$Q_4 = 2$$

$$R_4 = 2$$

INDEX

Absorbancy, 129
Acceleration, 326
Accumulator, 241
Address, 242
Aleksandrov, A. D., 17
Anand, B. K., 300
Anderberg, M. R., 260
Antiderivative, 332
Antonius, M. A., 100
Approximation, 14, 228, 247, 323, 336
Arithmetic mean, *see* Average arithmetic
Assumptions, 12, 13
Atkins, G. L., 146
Atkinson, D. E., 275
Auslander, D. M., 16
Average, 61, 340
  arithmetic, 62, 340
  logarithmic, 342
  geometric, 342
  harmonic, 343
  weighted, 62
Average rate, 113, 116, 140

Bailey, N. T. J., 16
Bartlett, M. S., 228
Base of logarithms, 333
Batschelet, E., 16
Beck, J. S., 28
Beckner, M., 3, 14, 17, 309, 313
Bee foraging, 104, 138, 255
Beer's Law, 127, 143
Benzinger, T. H., 290, 293
Bertalanffy, L. von, 16
Binomial coefficient, 75
Binomial distribution, 80, 86, 88, 89, 91

Biological rhythms, 210
Bits, 241
Bligh, J., 293
Book-balancing equation, 29, 159
Booth, D. A., 296, 300
Bray, G. A., 293, 300
Brazier, M. A. B., 300
Brobeck, J. R., 300
Brody, B. A., 17
Bross, I. D. J., 16
Brownlee, K. A., 234
Bruley, H. R., 282
Buck, R. C., 309, 313
Built-in functions, 249
Bungay, D. F., 282

Campfield, L. A., 293, 300
Carroll, L., 18, 39, 240, 261
Carrying capacity, 175
Cartesian product, 71, 314
Casey, E. J., 285, 293
Causal relation, 35
Chance, B., 275
Characters, taxonomic, 258
Charney, J., 194
Chemical catalysis, 267
Chemical reaction, 171, 267
  biomolecular, 133
  rate of, 135
  reaction event, 134
  reaction probability, 133
  unimolecular, 123
Chemostat, 52, 114, 276
Clark, L. R., 139, 164
Classification, 258

Classification hierarchy, 259
Clifford, H. T., 260
Closed system, 172
Clow, D. J., 16
Clustering *see* Classification
COBOL, 246
Cochran, W. G., 213
Code, C. F., 300
Coefficient of determination, 236
Coefficient of variation, 91, 96, 169
Coexistence of prey-predator, 304
Combinations, 71
Compartmental models, 28, 139, 159, 269
Competition, 177, 183
 direct, 173, 302
Competition event, 173
Component, 19, 27
Component diagram, 22, 33
Compiler, 245
Computer, 28, 240
Computer operations, 241
Concentration, 126, 151, 162
Conditional probability, 82
Conrad, H. R., 300
Conservation law, 30
Constant rate process, 103
Continuous system, 164
Controller:
 On-off, 285, 295
 proportional, 284, 287
Conversion factor, 41, 47
 approximate, 48, 55
Cooper, C. F., 9
Cooperativity, 141, 176, 183
Correlation, 234
 forced, 236
Correlation coefficient, 236
Correlation ratio, 236
Corresponding points, 56
Corresponding times, 139
Counting rules, 71
Cowgill, G. R., 55
Craster, E., 145
Crow, J. F., 16
Curnow, R. N., 37
Cyclic behavior, 184, 206, 210
 chemostat, 281, 284
 enzyme system, 268, 274
 prey-predator system, 210, 305
 synchronous populations, 140

Data transformation, 232
DDT, 28
Decomposition, 28
Decomposition of systems, 18, 27
Decreasing sensitivity, 142, 182
Degrees of freedom, 215, 226, 229
Delay, *see* Time lag
Delbruck, M., 10
Density, 126
Dependency statements, 21
Derivatives, 324, 331
 chain rule for, 328
 dimension of, 42
 formulas for, 327
 partial, 327
 second, 326
Deterministic models, 96, 102, 110
Deviations, random vs. systematic, 226
Diagrams, *see* Component diagram; Digraph; Graph; Input-output diagram; Signal-flow graph; State space diagram; Symbol-arrow graph; *and* Tree
Diamond, J. M., 60
Difference equations, 140
Differential equations, computation, 252
Diffusion
 continuous, 164
 discontinuous, 124, 138, 146
Diffusion coefficient, 166, 171
Digraph, 32
Dimension, 36, 39
 definition of, 45
 of state space, 101
 of variables, 21
Dimensional homogeneity, 49
Dimensionless ratio, 41, 53, 58
Diminishing returns, 142
Direction in state space, 158, 204, 207, 274
Discrete state space, 102
Disjoint events, 66, 69, 84
Disjoint sets, 311
Distance, 216
 Euclidean, 217, 257
DNA, 10
Doubling time, 122
Dowd, J. E., 238
Drabkin, D. L., 55
Dynamic process, 100, 145
Dynamic similarity, 58
Dynamic system, 58, 100

INDEX

Economic demand, 182
Ecosystem, 52
Einstein, A., 61
Ellis, B., 60
Elsasser, W. M., 17
Empty set, *see* Null set
End-point in state space, 155, 184
  moving, 155, 189
Environment of system, 138, 146, 157, 160, 189, 203, 207, 301
Enzyme control, 267
Enzyme reactions, 178
Equilibrium, 151, 155, 187
  chemical, 172
Equilibrium constant, 173
Errors
  expected, 215, 225, 228
  experimental, 6
  measurement, 98, 230
  rounding, 247, 252
  standard, 226
  structure, 232
Expected rate, *see* Average rate
Expectation, 63, 64, 85, 96, 203
Expected value, *see* Expectation
Extrapolation, dangers of, 7
Extrema, 219
Events, 68, 84, 92
Events, elementary, 112, 144

Falconer, D. S., 16, 96
Feedback, 33, 185, 189, 203, 211
  negative, 33, 193, 304
  positive, 33, 194
Feller, W., 74
Fencl, Z., 282
Flow, 28
Food intake control, 293
FORTRAN, 246
Fractional life, 121
Free, J. B., 103
Frenkel, R., 268, 275
Fructose diphosphate, 268
Functions, 318
  continuous, 323
Fundamental dimensions, 40

Gallucci, V. F., 188
Garfinkel, D., 16, 275
Garfinkel, L., 275

Genetics, 64
  Mendelian, 64, 94
  pea plant, 79, 90, 94, 97, 254
Geometric similarity, 56
Gessner, U., 293
Ghosh, A. K., 275
Glucose, 295
Glycolysis, 268
Goel, N. S., 62
Gold, H. J., 123, 135
Goldman, S., 200
Goodness-of-fit, 215, 226
Gordon, G., 250
Grabe, C. A., 16
Graph, 32
  connected, 259
Gray Larch budmoth, 139
Green, S. B., 275
Greenspan, D., 247
Grossman, S. I., 16, 140
Growth rate, 114, 142

Hale, J. F., 16
Half-life, 121
Hall, R. C., 203
Hamilton, C. L., 300
Hammel, H. T., 293
Harrison, H. L., 28
Hartigan, J. A., 260
Heat capacity, 285
Heat distribution, 289
Heat production, 289
Heat transfer, 285, 291
Hein, P., 1, 27
Heinrich, R. H., 275
Hempel, C. G., 17
Hemker, H. C., 16
Hensel, H., 289, 290, 293
Herskowitz, I. H., 64
Hess, B., 16, 275
Higgins, J., 211, 268, 275
Hill equation, 182
Holling, C. S., 307
Holwill, M. E., 285, 293
Homeostasis, 193
Homeostatic index, 199, 202, 203, 281, 304
Homogeneity with respect to independent variable, 120
Homogeneous population, 115
Horowitz, S. L., 228

Howland, J. L., 16
Hull, D. L., 17, 309, 313
Hulme, E., 275
Hunger Center, 295

Identical stochastically, 114, 117, 137
Increasing sensitivity, 140, 183
Independence, stochastic, 78, 83
 assumption of, 114, 117, 129, 137, 169
Infinite series, 338
Initial value, 119
Input, 19, 28, 97
 net, 19, 31
Input-output diagram, 23, 24
Input-output relation, 23, 28, 31, 34, 98
Integral, 329
 dimension of, 42
Interaction, 19
 direct, 133, 176, 200
 indirect, 148, 178
Interaction event, 137
Interpolation, dangers of, 6
Intersection of events, 69, 77
Intersection of sets, 311
Intrinsic rate of increase, 176, 201, 302
Intuition, 15, 144
Isolated system, 146, 153, 187
Iterative procedure, 248

Jacquez, J. A., 164
Jardine, N., 260
Jinks, J. L., 16, 96
Joint probability, 78

Karlin, S., 62, 108
Kimura, M., 16
Klir, G. J., 16
Komogorov, A. N., 17
Koong, L. J., 300
Koshland, D. E., 275
Kowal, N. E., 203
Kuhn, T. S., 7

Label techniques, 163
Lambert-Beer Law, 129
Langhaar, H. L., 60
Lavrent'ev, M. A., 17
Leaf canopy, 143
Least squares, 218, 228
Least squares weighted, 231

Le Megnen, J., 293, 300
Lepkovsky, S., 300
Lewis, W. M., 57
Light absorption, 127, 143
Likelihood, 68
Limit, 111, 321, 323, 325, 329
Limit cycle, 207
Limited resource, 178
Linear approximation, 339
Linear equation, parameter estimation, 219
Linear rate laws, 114, 117, 120, 169, 184
Liston, J., 260
"Little oh," 336
Lockhart, W. R., 260
Logarithms, 41, 332
Logistic curve, 175
Lotka, A. J., 210
Lotka-Volterra model, 200, 306
Lower order of magnitude, 336
Lucas, A., 268, 275
Lucas, H. L., 16, 300

Machine language, 245
Macko, D., 16
Macon, N., 247
Maleck, I., 282
Marcus, R., 246
Mark-recapture, 164
Marynick, D. S., 192
Mass action, law of, 133
Mass and weight, 50
Mather, K., 16, 96
Maximum, 220
McCracken, D. D., 246
McMahone, J., 59
Meadows, D. H., 194
Mean square, 226
Menke, W. W., 255
Measurement scale, 45
Memory in dynamic system, 101
Mental image, 144
Mesarovic, M. D., 16
Michaelis constant, 181
Michaelis-Menten, 278
Michaelis-Menten mechanism, 178, 278
Microorganisms, continuous culture, 275
Milhorn, 16
Miller, D. R., 203
Miller, J. G., 16
Milsum, J. H., 16

INDEX

Minimum, 220
Mobility, 166, 171
Model, building protocol, 25
Model
  correlative, 3, 34
  definition of, 2
  deterministic, see Deterministic models
  explanatory, 8, 18, 102
  mathematical, 2
  non-uniqueness, 6, 7, 9, 13, 38
  stochastic, see Stochastic models
Model testing, 99, 226
Molecular motion, 124
Moore, R. E., 293
Morowitz, H. J., 60
Multinomial distribution, 81
Murphy, C. F., 258
Mutually exclusive, see Disjoint events

n-dimensional space, 101
Newman, J. R., 17
Neyman, J., 228
Nonhomogeneity with respect to independent variable, 137
Nonhomogeneous population, 129
Novick, A., 282
Null set, 311

Observability in steady state, 192
Olby, R., 10
Operational taxonomic units, 258
Open loop gain, 199
Open system, 155, 157, 187
Ordered pair, 314
Ordered n-tuple, 314
Oscillations, see Cyclic behavior
Outcomes, 66, 68, 85, 92
Output, 19, 28, 96
  net, 19, 31

Pankhurst, R. C., 40, 60
Parameters, 31, 97, 213
  sensitivity, 203
  time dependent, 138
Parks, J. S., 142
Partition, 84, 259, 313
Pattee, H. H., 16, 17
Patten, B. C., 307
Pavlidis, T., 228
Peeling-off of exponentials, 131

Pest control, 255
Phenogram, 260
Phosphofructokinase, 267
Pielou, E. C., 16, 210, 306, 307
PL/1, 246
Pollard, E. C., 141
Poisson distribution, 108
Population dynamics, 173, 306
Prey-predator system, 30, 177, 190, 300
  Lotka-Volterra model, 201, 207
  Simulation, 252
Pring, M., 275
Pritsker, A. A. B., 250
Probability, 42, 63, 69, 92, 255
Probabilistic models, see Stochastic models
Probability distribution, 80
Programming languages, 246
Proof, mathematical versus physical, 12
Public address system, 18, 42, 97, 138, 191, 194
"pure death" process, 123, 125
Pye, E. K., 275
Pythagorean theorem, 216

Qualitative behavior, 5
Qualitative descriptions, 19

Rabins, M. J., 16
Radiation damage, 138, 141
Radioactive decay, 124
Random migration, 124, 164
Random number generator, 254
Rangazas, G., 268, 275
Range of a function, 318
Rape flower, 103
Rapoport, S. M., 275
Rapoport, T. A., 275
Rashevsky, N., 16
Rate of change, 326
Rate constant, 120
Realization of stochastic model, 255
Rearrangements, 71
Regression, 219
Relations between sets, 313
Relative frequency, 63
Relaxation time, 121, 146
Rensch, B., 17
Rescigno, A., 28, 307
Residuals, 227
Richter-Dyn, N., 62

Richardson, I. W., 307
Riggs, D. S., 12, 16, 33, 55, 60, 200, 237, 239, 240, 293
Roberts, F. S., 16
Root mean square, 226
Rosen, R., 16
Rubinow, S. I., 16
"rule of seventy," 122
Russell, B., 17

Sample space, 67, 82, 92
  continuous *versus* discrete, 85
  equal likelihood, 67
Sampling, 72
Satiety center, 295
Saturation effect, 142, 178, 183
Saturation of carrying capacity, 176
Scale, 102. *See also* Spatial Scale; Time scale
Scale factor, 56
Scale-independent ratio, 52, 58
Scale limitations, 59
Scale model, 56
Scale for plotting, 238
Scaling, 58
Schrodinger, E., 10
Sensitivity analysis, 197, 203
Separatrix, 205
Series approximation, 338
Series expansion, 248
Set-point, 290
Sets, 308
  compliment, 312
  equality, 310
  fuzzy, 309
  membership, 309
  polytypic, 309
  universal, 312
  see also Disjoint sets; Intersection of sets; Subset; *and* Union of sets
Shape factor, 55, 57, 340
Shaw, G. B., 186
Shipley, R. A., 164
Sibson, R., 260
Sigmoid curve, 183
Signal-flow graph, 21, 24, 32, 144
Signal-to-noise ratio, 236, 237
Signal splitter, 23
Signed digraph, *see* Symbol-arrow Graph
Similarity, 39, 51, 58, 257

Simplification, 146
Simulation, 250
  continuous system, 250
  discrete event, 253
  general purpose programs, 250
  Monte Carlo, 255
Smeach, S., 123, 135
Smith, J. M., 16, 211
Sneath, P. H. A., 258, 260
Snedecor, G. W., 213
Sokal, R. R., 256, 260
Southwood, T. R. E., 164
Spatial scale, 170
Stability, 191, 193
Stadtman, E. R., 268, 275
Standard deviation, 89
Stark, P. A., 247
State, 31, 96
State space, 43, 58, 100, 107, 157, 203
  continuous *versus* discrete, 102
State space diagram, 153
State variable, *see* Variables
Stationary, wide sense, 212
Stationary probability distribution, 95
Statistics, 213
Steady state, 157, 186
  chemostat, 281
  and cyclic behavior, 184, 211
  enzyme system, 179, 273
  and observability, 192
  population models, 174, 190, 201, 302
  sensitivity of, 203
Steel, R. G. D., 213
Stephenson, W., 260
Stochastic models, 36, 96, 102, 106, 135
Stoichiometric coefficient, 136
Stone, P. H., 194
Structural stability, 98, 306
Subpopulation, 75, 129
Subset, 68, 84, 92, 310
Subsystem, 18, 19, 27, 34
Summer of signals, 23
Surface area, 54, 291, 340
Sylvester, N. R., 285, 293
Symbol-arrow graph, 33, 144
Synchronized population, 107, 139
Synergism, 177
System behavior, 24
System equations, uncertainty in, 98
Szilard, L., 282

INDEX

Takahashi, Y., 16
Talbot, S. A., 293
Tangent line, 324
Taxonomy, *see* Classification
Temperature, dimension, 46
Temperature control, 282, 295
Tempest, D. W., 282
Tepperman, J., 300
Theory, nature of scientific, 10
Thermostat, 52, 282
Thingstad, T. T., 282
Threshold, 183
Time curve, 150, 155
Time lag, 23, 101, 144, 211
Time scale, 112, 140
Time scale separation of variables, 146, 270
Timofeeff-Ressovsky, N. W., 10
Titman, D., 182
Toates, F. M., 296, 300
Torrie, J. H., 213
Total probability, formula for, 84, 94, 107
Tracers, 163
Trajectory in state space, 100, 206
Transfer between compartments, 30, 159
Tree, type of graph, 259
Turner, J. E., 16, 140

Uncertainty, 96, 103, 104, 115. *See also* Errors
Union of events, 69, 93
Union of sets, 310

Units of measurement, 40
Urquhart, N. S., 16

Value of a function, 319
Variables, 20, 31, 43, 100
   control, 276
Variance, 88
Vector, 316
Velocity, 326
Volterra, V., 210
von Foerster, H., 200

Wall, A., 228
Watt, K. E. F., 16, 307
Weidhaas, D. E., 203
Weight and mass, 50
Weighted mean, 62, 231
Weights, likelihood, 93
Weinberg, G., 246
Wellington, W. G., 106
Whittow, G. C., 293
Williams, E. J., 213, 228
Williams, F. M., 282

Yasukawa, N., 246
Young, T. B., 282

Zadeh, L. A., 309, 313
Zar, J. H., 213
Zimmer, K. G., 10